Planning Agricultural Research: A Sourcebook

Planning Agricultural Research: A Sourcebook

Edited by
Govert Gijsbers
Willem Janssen
Helen Hambly Odame
and
Gerdien Meijerink

International Service for National Agricultural Research (ISNAR)
The Hague
The Netherlands

CABI *Publishing*
in association with the
International Service for National Agricultural Research

CABI *Publishing* is a division of CAB *International*

CABI Publishing
CAB International
Wallingford
Oxon OX10 8DE
UK

Tel: +44 (0)1491 832111
Fax: +44 (0)1491 833508
Email: cabi@cabi.org
Web site:http://www.cabi.org

CABI Publishing
10 E 40th Street
Suite 3203
New York, NY 10016
USA

Tel: +1 (212) 481 7018
Fax: +1 212 686 7993
Email: cabi-nao@cabi.org

A catalogue record for this book is available from the British Library, London, UK.

A catalogue record for this book is available from the Library of Congress, Washington DC, USA

Published in association with:
International Service for National Agricultural Research (ISNAR)
PO Box 93375
2509 AJ The Hague
The Netherlands

Tel: (31) 70 349 6100
Fax: (31) 70 381 9677
Email: isnar@cgiar.org
www.cgiar.org/isnar/

ISBN 0 85199 401 6

Text editor: Michelle Luijben-Marks.
Front cover design by Richard Claase.

Printed and bound in the UK by Cromwell Press, Trowbridge.

Contents

Foreword

A reassessment and reappraisal of the usefulness of planning has been underway since the disappearance of the centrally planned economies. The very role of planning has been questioned, as the "market knows best" view gained ground. To some, planning and plans have come to be seen as relics from a time gone by. They might argue that plans are often obstacles rather than useful tools in change processes.

Yet new approaches to planning are emerging that emphasize the use of plans to identify strategic issues and to help organizations adjust to rapidly changing conditions in the external environment. These ideas are relevant for agricultural research organizations, which must balance the need to adjust to changing circumstances with the long-term nature of agricultural research. Typically, agricultural research programs tie up resources for many years.

Throughout its 20 years of existence ISNAR has worked in areas related to planning agricultural research. Efforts have involved the development of methods, tools, and procedures, as well as their application through collaborative work with partner organizations.

This *Planning Sourcebook* builds on recent literature and on ISNAR's research and experience to discuss key developments in the context of research planning, the content of plans, the process of planning, and the tools that support planning efforts. The book brings together some 35 authors from different institutions and backgrounds providing a variety of planning perspectives.

The book was written especially for research managers and planners in agricultural research and technology organizations. We hope that the new insights and views on planning will be of use in their work and will help in the preparation and implementation of plans that enhance both organizational performance and accountability to key stakeholders.

Stein W. Bie
ISNAR, Director General

Acknowledgements

In this sourcebook we attempted to present a variety of different perspectives on agricultural research planning, based on contributions by a broad range of authors from different types of institutions. Contributions from 36 authors, representing more than a dozen organizations, have added richness to the texture of experiences captured in this book. We express our gratitude to those authors.

We appreciate the support that Stein Bie, ISNAR Director General, gave throughout this sourcebook's preparation. Howard Elliott and Francis Idachaba were also instrumental in providing help and useful comments and suggestions.

Many people helped us turn a series of papers into this book. Jan van Dongen, head of ISNAR's publication unit, coordinated the production. Michelle Luijben-Marks did the editing and designed the book's layout. Monica Allmand, head of ISNAR's library helped track down a large number of references. Rivka Peyra did the final proofing, Richard Claase designed the cover, and Claudia Forero provided secretarial support. We express our heartfelt thanks to all.

Finally, we extend our appreciation to the four reviewers who took the time to read the manuscript and make important comments and suggestions.

The editors

Introduction

Govert Gijsbers

Why a planning sourcebook?

Agricultural research is an investment in future production, productivity, and food security. But it's an uncertain business, because the investments required are large and benefits are uncertain and far away. Planning in agricultural research aims to guide investments towards the most relevant outputs in the most cost-effective manner. Research planning can be undertaken in a number of ways. Broad strategic issues may be the focus of planning work, but operational management also needs to be addressed. Planning may be action-oriented, or it may be directed towards the joint development of a shared vision on the future. Planning may be highly systematic, following established methods and procedures. Or it may be more erratic and ad hoc. But in the end, every research organization does planning, either explicitly or by default, and thinking on ways to improve planning continues to evolve. This book aims to provide insights, procedures, and tools that help managers establish planning approaches and procedures best suited to the conditions of a particular organization.

Evolution of approaches

Over the years, large amounts of money have been spent developing agricultural research plans. Planning's earlier popularity was due to the fact that it fitted neatly into a development model in which the public sector played a key role in agricultural development. Since the demise of the centrally planned economies around 1990, however, market-friendly forms of economic organization have become dominant; and planning in general has lost some of its legitimacy and credibility. In addition, the feasibility of research planning, particularly the feasibility of planning 10–15 years ahead has become doubtful in an increasingly dynamic environment. How amenable is the research process – in which creativity, flexibility, and serendipity play major roles – to rigorous scheduling of activities?

As a result, serious questions are being asked about planning. Some management gurus, such as Peters (1989), propose that organizations should learn to "thrive on chaos" rather than on planned interventions. Others, such as Mintzberg (1994), emphasize the need for strategic thinking, arguing that traditional planning models may not be the best way to arrive at strategies. He sees administrative and bureaucratic procedures as major obstacles in the creative search for new opportunities. Yet others maintain a positive outlook on planning, arguing that a revival of strategic planning is underway both in the private sector (Galagan 1997) and in the public sector (Bryson 1995).

There are several reasons why planning, in one form or another, will continue to be key for management. First, in a dynamic external environment it is imperative for all organizations to revisit strategic decisions frequently, and some process is required to arrive at and communicate strategies. Second, decisions must be made about long-term investments, and these will constrain future activities for many years. Third, and particularly relevant for agricultural research, external funding agencies insist on a sensible investment plan before making funds available.

Criticism of traditional planning models has led to the development of new approaches to planning. This book integrates many of those ideas and approaches in the materials presented. Table 1 presents a stylized summary of some of the main differences between "traditional" planning models and more recent approaches. There is considerable overlap between the different dimensions (represented in the rows); and the difference between traditional and more recent approaches is not absolute.

Planning used to be highly centralized, top-down, and technocratic. It was essentially an exercise done by senior management, in some cases assisted by a staff of professional planners, but with limited participation from outside the executive circle and without inputs from key external stakeholders such as clients and financiers. Now, there is general recognition that plans need to be relevant to the concerns of clients, and a broad section of the organization's staff should be involved in their making.

The centralization of planning led to a situation where many organizations or units were simply instructed by higher authorities to develop a plan. Compliance with administrative procedures dictated development of plans as a "a cal-

Table 1. Planning Approaches Compared

Traditional Planning Models	Recent Planning Approaches
centralized	decentralized
hierarchical	participatory
compliance	entrepreneurship
resource-based	objective-driven
internally focused	external orientation
long time horizons	shortening time horizons
implementation ignored	implementation a key issue

endar-driven ritual." Once the plan was approved at the highest level it became an operational objective. But in recent approaches to planning a more entrepreneurial spirit has gained the upper hand. Today, planning is seen as an instrument that helps organizations identify their niche and objectives vis-à-vis competitors and become more productive and relevant to stakeholders' needs.

The driving force behind plans and strategies has thus undergone a major shift, from planning models based on internal resources and capabilities towards models that emphasize objectives derived from analyses of the external environment. While internal resources and mechanisms are relatively constant (if not rigid), many organizations' external environment is changing at an unprecedented rate – as the pace of competition speeds up and as new technologies change the ways in which business is done. These days, a major challenge for any organization is to identify its "niche" in a landscape where competitors and partners have their own territorial ambitions.

In the process, time horizons for planning have become shorter. Organizations now realize that it is next to impossible to make detailed, quantitative forecasts for a 10–15 year period. Projections have given way to more indicative forms of planning. For the longer term, say 10 years, the emphasis is now on qualitatively different strategies or scenarios, rather than on quantitative targets to be achieved. This creates, of course, problems for long-term finance and investment planning, where decisions taken now have implications far beyond any feasible "planning horizon."

Finally, awareness has grown of the key role of plan implementation. In traditional planning, implementation was not much of an issue; it was assumed to be a straightforward administrative process to be undertaken by operational staff under instructions from decision makers higher up in the organization. But now implementation is seen as an urgent problem and a very difficult task – witness the large number of plans that gather dust on bookshelves throughout the world. Implementation is now considered a complex, political process that requires leadership and management, as well as feedback through careful monitoring and evaluation.

The emergence of new approaches to planning does not mean that the more traditional models have become completely obsolete. Whether to use a traditional or more innovative approach depends on a number of factors. The newer approaches are particularly useful when strategic reorientation in a dynamic external environment is an issue. Traditional approaches are useful in operational types of planning where the broader strategic framework is already in place.

For agricultural research organizations and systems, planning faces a number of specific challenges. Agricultural research organizations are under pressure to demonstrate to national and international financiers that they provide value for money. They must give evidence of their performance and of their relevance to stakeholders' needs. Increasingly, agricultural research organizations function in a context of blurring boundaries between the public and private sectors, underscoring the importance of a clear strategic outlook. They need to deal with new funding mechanisms and find answers to the challenges presented by new technologies.

This *Planning Sourcebook* is an attempt to contribute to better understanding of the many aspects of agricultural research planning, particularly, to present and analyze recent developments and future challenges. An overview of different aspects of agricultural research planning is presented through a series of chapters that give specific examples of the different types, approaches, and styles of research planning. A glossary at the end of the book provides a quick reference on planning terminology.

Target audience

Practitioners rather than academia are envisaged as the main audience for this sourcebook. It will be particularly useful for those working in agricultural research as planners, research managers, and directors at the research program or organization level. The book will also serve the needs of researchers wanting to gain an understanding of planning. Finally, the book will be useful for staff at international and donor organizations who deal with aspects of planning agricultural research.

Organization of the book

This *Planning Sourcebook* consists of five parts. Parts I, II, and III, following de Wit and Meyer (1998), deal with the context of planning, the content of planning, and with planning processes. Part I, on context, addresses the *where* question of the external environment in which agricultural research planning is done. Part II, on content, deals with the *what* of planning, discussing the substance of different types of planning. Part III, on process, looks at the *how, who*, and *when* of planning: how it is organized, who is responsible for certain tasks, and in what manner different tasks and plans are scheduled to fit financial and agricultural cycles. Part IV examines the use of various analytical tools that support planning processes. Part V provides additional information in the form of a glossary and other information sources.

Context, Content, Processes, and Tools

Each chapter in parts I through IV begins with an overview of the main issues and problems related to the topic confronting practitioners. These introductions are followed by the chapters, which detail specific subjects, types, topics, tools, and issues in agricultural research planning. The chapters are written to a standard outline to help ensure they are easy to use and relevant to practitioners.

1. **Summary.** Briefly summarizes the paper.
2. **What is . . . ?** Presents definitions, objectives, and purpose of the method or tool being described (e.g., "What is strategic planning?").
3. **How to . . . ?** Shows how to use, do, or deal with the topic (e.g., "using logical frameworks" and "doing master planning").
4. **Usefulness or relevance for agricultural research organizations.** Assesses the importance of the topic for decision makers in research organizations.
5. **Examples.** Contains case studies or illustrations from agricultural research planning practice.
6. **References, recommended reading, other sources of information.** Presents key literature references and other relevant information such as websites.

Annexes

Part V contains a glossary of planning terms covering a broader range of concepts than those included in the information digests. Further, an annotated bibliography presents the a selection of important publications in the field. To help readers find additional information on the Internet, a website section contains references to ISNAR's website, where relevant materials are available, as well as to other important sites and pages.

References

Bryson J. M. 1995. Strategic Planning for Public and Nonprofit Organizations – A Guide to Strengthening and Sustaining Organizational Achievement. San Francisco: Jossey-Bass.

De Wit, B. and R. Meyer. 1998. Strategy: Process, Content, Context – An International Perspective. London, ITP.

Galagan, P. A. 1997. Strategic planning is back. *Training & Development,* 32–7.

Mintzberg, H. 1994. The Rise and Fall of Strategic Planning. New York: Prentice Hall.

Peters, T. 1989. Thriving on Chaos: Handbook for a Management Revolution. Harper Collins Publishers.

Part I
The Context of Agricultural Research Planning

Willem Janssen

The speed of change in many societies appears to be accelerating, now as in the past decade, as agricultural research systems throughout the world have faced new challenges and prospects. Many countries have reorganized their research systems in line with structural adjustment policies. Some have invested heavily in biotechnology or emphasized work in natural resource management. Others have yet to decide which direction their research effort should turn. To understand how the context of agricultural research is changing, it may be helpful to distinguish three main categories of change (Janssen 1999) (see figure 1):

1. changes in the demand for agricultural research and knowledge
2. changes in the supply of research and knowledge, within and outside agriculture
3. changes in the organization of the public sector and conditions for public and private management

As an introduction to the chapters on the *context* of agricultural research planning, this section presents an overview of developments within each of these

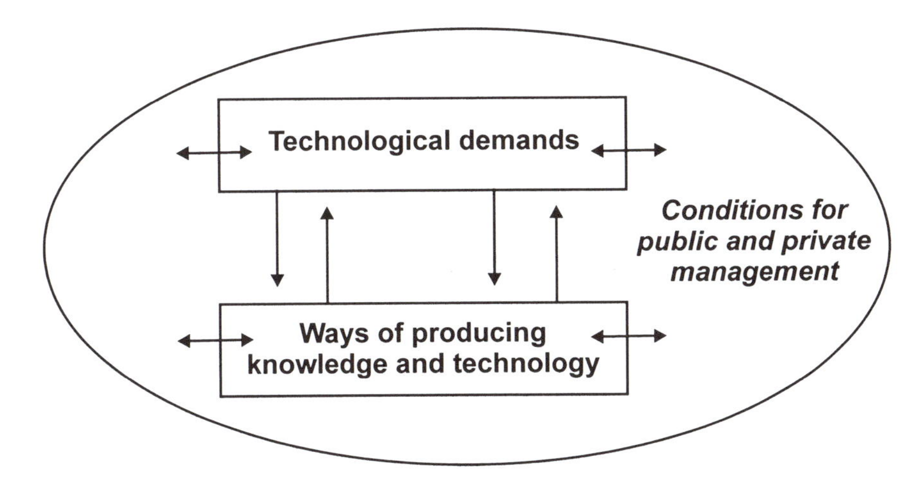

Figure 1. Categorization of the major changes in the context of agricultural research

categories and their implications for agricultural research planning. Rather than dwelling on experiences in specific countries, the description focuses on broad issues that are relevant in most countries.

Demand for research and knowledge

Demand has grown fast for technologies to integrate agriculture and *environmental concerns*. In the developed countries, worry about sufficiency of food has diminished due to past agricultural research success and to low food prices induced by subsidies. Moreover, urban populations are starting to realize that rural areas fulfil functions other than agriculture, such as water supply, recreation, and nature conservation. In developing countries, it has become evident that growth of agricultural production without improved natural resource management leads to resource degradation. Alongside safeguarding future harvests, such negative external effects have to be avoided. For example, (agriculturally induced) soil erosion not only reduces the future production potential of land. It also affects people who depend on fishing for their livelihoods and may cause flooding downstream. Resource management technologies are thus increasingly seen as central to achieving sustainable production increases (ISNAR 1998).

A very different demand for agricultural research stems from the need to raise *competitiveness*. In the process of globalization, opportunities for agricultural development may be found in specializing and supplying external markets. Competitive advantage in these external markets depends on factors other than just the cost of supply. Product quality, timely delivery, and ability to supply appropriate amounts are central themes. Market information systems and market-oriented research become more important. Competitiveness matters in domestic markets too, when internal markets are opened to external suppliers. Traditionally, most research focused on yield increases or cost reduction. But to raise competitiveness, the whole supply chain must be examined, including producers, traders, processors, retailers, and other chain actors (ISNAR 1998).

Supply of research and knowledge

Agricultural biotechnology, which generates and applies knowledge about the molecular structure of the genome, offers opportunities to overcome constraints that cannot be addressed using traditional research methods. It may also improve the efficiency of some traditional methods of research. Especially in the more developed countries, agricultural biotechnology is entering the mainstream of technology generation. Genetic markers and tissue-culture techniques are also gaining ground in developing countries' agricultural research systems (ISNAR-IITA 1999). Yet agricultural biotechnology is only just starting to have an impact on poor consumers and producers.

Developments in the field of *information* are making it easy to gain information made available elsewhere in the world. The Internet provides access to databases and the ability to process large volumes of information has also increased, leading to greater emphasis on modeling and database management (e.g., geographic information systems; see Pachico, this volume). On the other hand, new information is often copyright-protected in order to reward the inventors for their efforts. While more information is available and accessible than ever before, it is available and accessible mainly for those who are able to pay for it. Thus, the advantages of the information era have not benefited everyone in the same way. While the Internet has dramatically improved access to information in the developed world, in the rural parts of many developing countries it is still mainly fiction.

Conditions for public and private management

Public and private responsibilities in society have been a constant topic of discussion during the last decade (Rappert 1995, Fuglie et al. 1996). The resource stringency of structural adjustment has led many governments to focus on "core" public responsibilities, leaving other tasks to the private sector. Public management specialists argue that basic research (for which the benefits are hard to predict and to appropriate) as well as research for poor, unorganized producers (who cannot afford to set up their own research effort) are public responsibilities. Commercial farmers, possibly through producers' associations, and agroindustries should fund and commission their own applied and adaptive research, according to these specialists. These ideas are discussed – and challenged – at theoretical and practical levels in most developing countries. The public-private balance in research will remain an important issue in the coming decade (Pray and Umali-Deininger 1998).

Emphasis on *public accountability* has also increased. While the public sector cannot be expected to be profitable, it should be efficient and responsible in its use of public funds. New public management theory has highlighted the need to define goals and objectives for public-sector activities and to evaluate the public sector with respect to these goals (Mayne and Zapico-Goñi 1997). Accountability implies a shift from input-oriented towards output-oriented management, and many public-sector activities are organized in projects with clearly defined objectives, output, and activities (Osborne and Gaebler 1993).

Implications for agricultural research planning

These trends have not affected all people and all countries equally. Those isolated from international trade and technological change may face worse conditions than they did in the past. Some 830 million people were still food insecure in 1998 (FAO 1999). Agricultural research thus continues to face a double challenge: contributing to overall economic growth and targeting its efforts towards the improvement of food availability and income for the very poor. In a context of rapid change, research organizations must provide knowledge and technology that is relevant to the society around them; they must also capture any opportunities to address problems created in this dynamic context.

Planning for change is becoming a key issue. Because the external context is changing so quickly, research organizations that do not adapt lose their relevance fast. A successful organization is not only efficient, but responds agilely and appropriately to changing circumstances.

Planning processes should be linked to organizational characteristics. Organizations that are highly relevant to client needs (i.e., they are working on the right issues, using the right tools) should emphasize capacity building to increase their impact and efficiency. If they are relevant and efficient, their challenge is to develop strategies that enable them to remain in that position (e.g., by means of market recognition and forecasting and by raising standards of performance). Organizations that function efficiently but produce knowledge and services of low relevance need to reorient themselves by strategically oriented planning. Organizations with both low relevance and low efficiency risk being left alone to break down gradually. They face an urgent need to reinvent themselves. Figure 2 summarizes planning and management strategies that might be suitable for organizations at these different levels of relevance and efficiency.

Current emphasis in planning is moving away from resource and routine planning towards strategic planning, planning by objectives, and support for continual organizational innovation processes. Changes in an organization's external context often drive planning processes. Reassessing the organization's relevance in the light of the changing context is now becoming the pivotal element in most planning (see also de Souza Silva, this volume).

<table>
<tr><th colspan="2"></th><th colspan="2">Efficiency</th></tr>
<tr><th colspan="2"></th><th>Low</th><th>High</th></tr>
<tr><td rowspan="2">Relevance</td><td>Low</td><td>"Reinvent or die," systemic changes need to be planned</td><td>Strategic reorientation
Emphasis on strategy and programs</td></tr>
<tr><td>High</td><td>"Capacity building," focus on resource planning</td><td>Continuous improvement strategies</td></tr>
</table>

Figure 2. Planning and management strategies for organizations at different levels of efficiency and relevance

The chapters

Two chapters in this section concern developments in the international setting. The *globalization* chapter (chapter 1) combines changes in demand for research with changes in how knowledge is produced. It suggests ways to position research organizations in order to capture the maximum opportunities and benefits of the globalization process. Competitiveness is a major consideration in questions posed by globalization. There are also concerns that poor countries do not have the capacity to benefit from globalization (ISNAR 1997). The chapter on globalization also looks at some of the losers in the globalization process and suggests how to ensure that the least-developed countries will not be hurt.

Chapter 2 assesses the *regionalization* efforts observed in many parts of the world. Regionalization is one response to improved information and communication opportunities, as well as to calls for enhanced performance and accountability. The chapter describes various forms of regionalization and identifies success factors for pooling research efforts. Differences between planning for regional and national organizations are also discussed.

Chapter 3, on integrating *natural resource management (NRM)* issues into agricultural research agendas, discusses the background to increased demand for research services in the field of resource management. It discusses how NRM-oriented research is different from agricultural research and what implications this has for planning. The chapter emphasizes that agricultural research organizations can be more relevant and efficient if they incorporate NRM issues from the start in their planning processes.

Chapter 4 discusses the planning implications of *new technologies* for agricultural research. New technologies tend to require up-front investments and are often associated with high uncertainty. Specific tools are proposed for dealing with the uncertainty of benefits. New technologies are shown to have consequences not only in terms of the resources required, but also in terms of the subjects they enable research organizations to address, public acceptance of research, and legal and regulatory frameworks that need to be put in place.

The final chapter of this section elaborates on calls for increased *performance and accountability* in public agricultural research. It explores the background to these demands, examines the relationship between accountability and organizational performance, and discusses what is required to improve accountability within research organizations. The highlight is on the need to focus planning and management systems on outputs and outcomes.

References

FAO. 1999. The State of Food Insecurity in the World. Rome: Food and Agriculture Organization of the United Nations.

Fuglie, K., N. Ballenger, K. Day, C. Kotz, M. Ollinger, J. Reilly, U. Vasavada, and J. Yee. 1996. Agricultural Research and Development: Public and Private Investments under Alternative Markets and Institutions. Agricultural Economic Report No. 735. Washington, D.C.: USDA Economic Research Service.

ISNAR. 1997. Annual Report 1996. The Hague: International Service for National Agricultural Research.

ISNAR. 1998. New Technological Demands: The Methodological Framework for an INIA/BID/ISNAR project. The Hague: International Service for National Agricultural Research.

ISNAR and IITA. 1999. Biotechnology for African Crops. Study commissioned by the Rockefeller Foundation. Mimeo. The Hague: International Service for National Agricultural Research.

Janssen, W. 1999. Tendencias en la Organización y el Financiamiento de la Investigación Agrícola en los Países Desarrollados. Mimeo. Montevideo, Uruguay: PROCISUR.

Mayne, J. and E. Zapico-Goñi. 1997. Monitoring Performance in the Public Sector – Future Directions from International Experience. London: Transaction Publishers.

OECD. 1999. Modern Biotechnology and the OECD. Policy Brief. Paris, France: Organization for Economic Development and Cooperation.

Osborne, D. and T. Gaebler 1993. Reinventing Government: How the Entrepreneurial Spirit is Transforming the Public Sector. New York: Plume

Pray, C. and D. Umali-Deininger. 1998. The private sector in agricultural research systems: Will it fill the gap? *World Development*, 26 (6).

Rappert, B. 1995. Shifting notions on accountability in public and private sector research in the UK: Some central concerns. *Science and Public Policy,* 22 (6): 383–390.

Chapter 1
Globalization: Planning Agricultural Research in an Open Market Economy

Steven R. Tabor

Globalization is leading to integration and increased interdependence among nations. It is a disruptive process with winners and losers. Globalization puts pressure on agricultural research organizations to remain or become competitive in increasingly open markets for goods and services, including markets for research and development (R&D). The globalization process has important implications for key agricultural research concerns: economic growth, food security, poverty alleviation, and environmental conservation. National agricultural research systems (NARS) must analyze the globalization process and plan strategies for integration into a global agricultural R&D system. They must identify their niche and reposition themselves accordingly. This has major implications for planning and planners.

What is globalization?

Globalization is breaking down barriers between countries, cultures, and scientific communities at a pace that was hard to imagine just a few decades ago. Technological advances and policy convergence have led to what the IMF (1997) describes as world economic, political, social, and cultural integration, even though labor movements are still rather restricted. Globalization has increased linkages and interdependence of national economies, spawned the creation of supranational decision-making bodies, and contributed to the harmonization of economic policy, domestic laws, and institutions. Sachs and Warner (1995) identified the following main forces underlying the most recent phase of globalization:

- technological revolutions in shipping, aviation, automation, and telephony, which increased access to information and lowered communication costs
- trade and exchange-rate liberalization, which triggered a rapid rise in cross-border trade and investment
- the end of cold-war political tensions, which spurred political convergence towards democratic, market-friendly forms of economic organization

- development of global media and the spread of common languages and business practices, which helped create common cultural norms and expectations
- growth and proliferation of the multinational enterprise
- growing commitment of sovereign nation-states to international decision making on matters of global importance

For many developing countries, the process of globalization has been intimately linked with the adoption of market-oriented forms of economic management. The role of government in the productive sectors has been reduced and growth led by the private sector has come to be the norm. Government policies, rather than commanding the markets, have become subject to the discipline of the market, particularly as private-sector capital is now so very mobile.

Globalization generates both winners and losers. It is a disruptive process with immediate costs and benefits. Opportunities for growth increase as market opportunities emerge and resources flow to where they can be used most efficiently. This has particular relevance for developing nations because of the important role that agriculture plays in their economies and because agricultural growth has long been hampered by small domestic markets, restricted access to international markets, and a relative paucity of capital, technology, and effective institutions.

But many of the world's poorest nations hardly participate in global commerce and, with weak institutions and few apparent investment opportunities, they may well be destined to remain on the periphery of the dynamic global economy. This group of countries, and particularly the poorest people within these nations, risk being left behind by globalization. In addition, globalization often appears asymmetric: while developing countries may open their markets, large agricultural subsidies continue to exist in the European Union, Japan, and the United States. These in effect make it impossible for smallholder producers from the developing world to compete in the global market.

Even in developing nations that are reasonably well integrated into the global economy, the cost of adjusting to the new economic, political, and cultural setting may be very high. Whole branches of agriculture and industry may be ill prepared to face the rigors of international competition. Demand for natural resources is set to increase in the newly emerging competitive sectors, and this may strain the management of fragile natural resource endowments. Scores of public institutions, exposed to private-sector and international scrutiny, may find that they provide goods and services that do not live up to global expectations (Haggerd 1995).

New paths for meeting agricultural research goals

Regardless of how the costs and benefits of globalization are distributed, there can be no denying that globalization is proceeding at a rapid pace and that it is a major force to be reckoned with. Globalization is fundamentally changing the setting in which agriculture operates and, as a corollary to that, the needs and interests of the stakeholders that agricultural R&D facilities serve. Globalization changes the setting for agricultural research systems in a number of ways (see Bonte-Friedheim et al. 1997):

- Pressure to become (or remain) internationally competitive forces the agriculture sector to seek ways of innovating, augmenting demand for new and improved technology.
- Communication and transportation costs fall while access to information rises, creating scope for greater domestic and international specialization in the production and dissemination of new technology.
- As the interests of the private sector become more directly linked to technological progress, farmers' associations, agribusinesses, and other agencies play more active roles in managing the development and acquisition of agricultural technology.
- Growing multinational and private-sector agribusiness activity leads to steady growth in the stock of proprietary technology suitable for tropical agriculture. This, in turn, spurs the diffusion of intellectual property rights regimes that offer adequate proprietary rights to innovators of agricultural technology.
- Global performance benchmarks, or standards, for different facets of agricultural R&D activities emerge against which the performance of an agricultural research organization can be assessed.

What globalization does not do, however, is change the main public policy goals and objectives of agricultural research. For most countries, these megagoals relate to overall economic and social development and include, *inter alia*, economic growth, food security, poverty reduction, and environmental preservation. Globalization does change the ways in which these objectives can be met and also the relative priority that governments may need to assign to each.

Economic growth

Globalization ushers in increasing opportunities for farmers to obtain new technology and agricultural research services from sources other than domestic research institutions. As "accessible" external sources of agricultural research and technology proliferate, it becomes important for domestic agricultural research providers to identify their "niche," or area of core competence. In defining this growth-supporting niche, agricultural research providers should look for focus areas that (1) have the potential to boost domestic productivity growth in competitive global markets, (2) enable them to be a globally competitive supplier of new knowledge for the particular problems identified, and (3) complement and adapt research results garnered from other sources for local conditions. Conversely, research providers that find themselves working on "sunset" commodities are providing noncompetitive services (or are likely to be bypassed by a forthcoming wave of technological advance elsewhere) and need to restructure their operations. For countries facing similar constraints to agricultural growth, globalization may facilitate international cooperation in agricultural R&D (see Perrault, this volume), which, in turn, can lower research costs and speed up technology development cycles.

Food security

With globalization, countries tend to rely more on trade and income policies to buffer poor producers and consumers from food availability shocks than on market controls and direct public stockpiling and provision of foodstuffs. As specialization in agricultural R&D occurs, the farm sector may become increasingly reliant on global technology. If the flow of global technology is disrupted – for example, if a multinational hybrid seed supplier should fail – it may be difficult to restore domestic R&D capabilities or locate another suitable international source of technology supply. To protect the national population from such a breakdown in technology provision, agricultural research systems may need to define and maintain a certain amount of superfluous local R&D capacity in key areas.

Poverty reduction

Farmers in well endowed regions, with good access to market infrastructure, are the ones most likely to benefit from improvements in access to global markets, investment, and technology inflows. The poor will also benefit, albeit indirectly, through increased labor demand in farming and agroprocessing. But many small farmers and landless laborers in poorly endowed areas grow crops for their own consumption and have few links to the global economy. Globalization is unlikely to solve these producers' technology problems, particularly when the commodities they produce are not widely traded in international markets. For public-sector research organizations, the poverty reduction challenge is to identify and address cases in which carefully targeted development of new technology, together with complementary rural development investments, can support a meaningful improvement in nutrition and productivity of the rural poor. Such investments may enable poor communities to generate the resources required to integrate themselves into the more dynamic, globalized segments of the economy.

Environmental preservation

Where globalization inspires agricultural growth, it will also increase demand to draw upon the natural resource base. In some developing countries, this will either aggravate existing resource management concerns or place new pressures on fragile natural resource bases. Recent global trade accords have tended to link market access to the way in which agricultural products are produced. To retain market access, agricultural research is needed to ensure that products are produced in a way that is consistent with internationally agreed environmental guidelines. Numerous global fora have promoted more sustainable forms of agricultural development. This has raised national awareness and led to greater emphasis on environmentally friendly technology. Networking, visits, conferences, and other forms of international exchange have helped the research community identify options for addressing shared environmental management concerns (Cooper 1994).

Rebalancing public priorities

While growth, poverty reduction, food security, and environmental preservation are likely to remain important development priorities, globalization is likely to change the balance of public priorities in agricultural research away from productivity enhancement towards poverty reduction and environmental sustainability. This does not necessarily imply that productivity enhancement will become less important, only that globalization is likely to boost inflows of productivity-enhancing technology while stimulating private-sector interest in developing agricultural technology.

In the area of productivity improvement, globalization should inspire the public sector to focus on strategic and basic research questions, but in a highly selective manner. Adaptive research tasks will increasingly be conducted by or with the partnership of the private sector. Resource management and environmental research will play a greater role as will efforts to buttress food security by creating "backup" technology in gene banks and strategic commodity laboratories.

Responding to globalization issues

Planning processes exist to serve decision making. As the decision-making setting changes, planning processes will need to adjust to ensure that information provided is relevant to the current setting. As nations become more globally integrated, the focus of agricultural research planning changes in three important respects.

Tracking globalization itself. Planners need to track changes in the global setting for agriculture and agricultural research. Agricultural research planners especially need to keep abreast of developments in the main global commodity markets, international agreements that affect agricultural market access, emergence of trading blocs and common investment regulations, and innovations in transportation, communication, and automation that ease information exchange. Understanding the pace of globalization will help research leaders understand the scope of and limits to national R&D decision making.

Planning integration of NARS into global R&D. Research leaders need to develop plans to integrate national agricultural R&D into global R&D systems. On the policy front, establishment of an enabling regulatory environment will enhance the flow of technology across borders. Enacting an effective intellectual property rights regime for agricultural goods and processes is one of the immediate global integration challenges facing many developing nations. For research providers, operating across borders involves a host of institutional innovations that require careful planning. Developing mechanisms to buy and sell technology, to contract-in and contract-out services, to cooperate on tasks with researchers from other countries, to serve on cross-border technology decision-making bodies, and even to staff organizations with scientists from other nations requires careful planning, some experimentation, and a great deal of learning by doing. These processes will be eased if the research community has access to the "technological tools" that allow it to participate in global exchange, such as modern telecommunications, the Internet, and travel. As global information acquisition comes to play an ever more important role in the agricultural research process, strategizing for acquiring and disseminating such information within the research community itself becomes an important topic for planning.

Role positioning and niche finding. Globalization changes national research priorities. Certain tasks previously in the public domain may need to be left to the private sector. Some areas will receive more public emphasis and others less. Research agencies need to position themselves for evolving public and private research priorities. But besides finding an appropriate role, research entities also need to determine whether or not their services fill a meaningful niche in the market for agricultural technology. Over time, globalization will put more sources of agricultural technology and expertise at the disposal of the globalized farm community. With more technology available, and more "off-shore" sources of research expertise to choose from, agricultural research providers must constantly examine whether the services they provide satisfy a meaningful niche in the technology market. If the niche is likely to be better served by other providers, either other fields of specialization should be identified or structural changes considered within the research organization. Finding an appropriate R&D niche is a difficult and information-intensive matter. Many countries use technology foresight studies as a way to anticipate agricultural market prospects and technological options and trends, and so to identify particular niches for national R&D efforts (Walsh et al. 1991). In England and Japan, for example, delphi techniques have been used to draw industry expertise into forecasting where markets will head and to pinpoint where domestic technology efforts are required (Wright 1997).

Planning and choice making

The global environment changes quickly and certainly not with the orderliness or predictability of annual budgets and five-year investment plans. Opportunities open, and are sought, grasped, or missed. New sources of information arise, are capitalized on, and just as quickly become outdated and outmoded. With globalization, the pace of change accelerates, if, for no other reason, due to the force of competition that the world economy brings to bear. In a volatile, dynamic environment, agricultural research leaders need to maintain a modicum of flexibility, while marshaling the information they need to grasp opportunities and make hard decisions to ensure relevance, efficiency, and effectiveness of agricultural research in an ever more complex and contested agricultural R&D environment.

Where important decisions must be made but are difficult to anticipate, a planning entity must be prepared and quick to respond. Informing decision makers so they can reach a "suitable" or "reasonable" judgment becomes far more important than attempting to provide a comprehensive blueprint for all the research planning decisions that one would like to make.

Research planners need to acquire appropriate information. Markets can be tracked, technology requirements monitored, activities of other research bodies scrutinized, and the investment plans and interests of both public policymakers and the agribusiness community taken into consideration. Much of this information is either confidential or proprietary; few farmers or agroindustries publish their future investment plans. To glean this information, research planners need to invoke open, participatory planning and decision-making processes.

Relevance for agricultural research

Planning agricultural research has long been a preserve of technical scientists. Researchers have become research planners either by rising within their organizations to become administrators or by being delegated to participate in planning exercises. In the closed public-sector-led economy of the past, the main agricultural research challenge was to generate as much relevant technology as a country could afford. Since the research community had a virtual monopoly, scientists were vested with responsibility for developing plans to keep the technology "shelf" well stocked.

But in today's open private-sector-led economy there are many ways to stock the technology shelf, and national agricultural research providers have to respond to many more actors and be far more flexible and adept at positioning the domestic agricultural research effort to complement that which can be provided internationally. In this environment, what is needed for agricultural research planning are individuals who are not necessarily good scientists but who are good at spotting and exploiting opportunities to deliver technology in a competitive manner.

As the agricultural research endeavor becomes a matter of cross-border business, agricultural research planners will need the skills of multinational business planners. Expertise in business management and public policy will have to be combined with the temperament of a small-business entrepreneur and the communication ability of a refined diplomat. Operating in an open information-demanding environment, the planner will need to draw representatives from the scientific community, public policymakers, the agribusiness sector, the farm community, and nongovernmental organizations into the planning process. To keep business moving forward, the planner will need to respond quickly to requests for information to support decision making. In addition to staying abreast of changing agricultural market and technological developments, the research planner will need to help forge strategic alliances with research service providers from other countries and cultures, and work equally well with representatives of the public, private, and nongovernmental sectors.

Globalization is drawing national economies together, both to cooperate and to compete. The same is happening, albeit at a slower pace, in the world of agricultural technology generation and diffusion. But the direction of change is reasonably clear. New agricultural market opportunities will emerge; new pressures will be exerted on fragile natural resource bases. Some producers, regions, and countries will benefit while others will be bypassed in these processes. The private sector (both multinationals and domestic) in developing countries will come to play a more important role in technology development and acquisition, and the field of agricultural technology generation will become more crowded, contestable, and ultimately, efficient as trade in technology and agricultural R&D expertise accelerates.

Research systems need to adjust to changing public and private priorities and to do this in a way that ensures they maintain a niche in the R&D chain that is both relevant and rewarding. They need to track the globalization process itself, both to understand what it means for their nation's agriculture sector and to understand their position in the evolving R&D marketplace. By planning how to integrate their R&D system into the global R&D effort, research leaders will avoid being bypassed. Creating an enabling environment for global information acquisition and management is a necessity if research leaders are to identify what they can do best. Using market, technology, and stakeholder information to realign institutional priorities and to find institutional research niches will become more and more important.

But globalization also ushers in a fast-paced, rapidly changing decision-making environment. If planning information is to be useful, it must feed into windows of decision-making opportunity that open quickly and shut even faster. Making a "reasonable," well informed decision in the very short period when decision-making opportunities emerge is now far more important than having a comprehensive blueprint for research decision making five years hence.

Planning must also come to grips with flexibility, especially in a setting in which discoveries in one part of the world can quickly make redundant the centerpiece of a research program's work in another part of the world. Scanning and monitoring the aims and achievements of R&D competitors is urgent if unproductive effort is to be avoided.

Much of the information needed to make sensible R&D decisions is both proprietary and area specific. Planning processes that draw on knowledge of the market and understanding of private agroindustry, on the one hand, and the farm sector, on the other, are needed to guide decision making in the right direction.

Clearly a new breed of agricultural planner is now needed. Leading scientists, well trained in solving complex problems or building sophisticated research institutions, are unlikely to have the skills, knowledge, or temperament to support decision making in a more diverse, global R&D environment. Individuals with knowledge and experience in business, international affairs, public policy, and communications are needed to tap into the perhaps confidential or proprietary information on markets, technology, public priorities, and R&D competition. Generating such information and feeding it into quick-gestating decision-making processes requires a modicum of entrepreneurship and business acuity – traits quite different from those nurtured through years of careful application of scientific techniques.

References

Bonte-Friedheim, C., S. R. Tabor, and H. Tollini. 1997. Agriculture and globalization: The evolving role of agricultural research. In *The Globalization of Science: The Place of Agricultural Research* edited by C. Bonte-Friedheim and K. Sheridan. The Hague: International Service for National Agriculture Research.

Cooper, R. 1994. Environment and Resource Policies for the World Economy. Washington, D.C.: The Brookings Institute.

Haggerd, S. 1995. Developing Nations and the Politics of Global Integration. Washington, D.C.: The Brookings Institute.

IMF. 1997. World Economic Outlook: Globalization Opportunities and Challenges. Washington, D.C.: International Monetary Fund.

Sachs, J. D. and A. Warner. 1995. Economic reform and the process of global integration. *Brookings Papers on Economic Activity*, 1: 1–118.

Walsh, V., I. Galimberti, J. Gill, A. Richards, and Y. Sharma. 1991. The Globalisation of the Technology and the Economy: Implications for the Scientific and Technology Policy of the EC. FAST Occasional Papers 248. Brussels: Commission of the European Communities, Forecasting and Assessment in Science and Technology.

Wright, D. 1997. Turning foresight into action. *Science, Technology and Innovation*, 10 (5): 18–25.

Chapter 2
Regionalization of Agricultural Research: Implications for Planning

Paul T. Perrault

Regional agricultural research is research that is coordinated and shared among institutes from various countries within one region. Regionalization is most effective for tackling issues that cannot be dealt with efficiently on the national scene, due to lack of funds or intellectual resources or because their impact extends beyond national boundaries. Regional efforts are often difficult to organize and maintain, however. Participants may be unable to assume full ownership of regional research agendas, or a regional body may lack the political oversight to address the issues submitted to it by members. While regionalization may improve the efficiency of research in a given region, its costs and benefits should be carefully weighed to ensure sustainability.

What is regionalization?

This chapter defines regionalization of agricultural research as transnationally organized or coordinated research that involves entities from a number of countries within a region (Gijsbers and Contant 1996). The problems that agricultural research aims to solve are not confined within certain national borders. Rather, they are often spread over several countries. For example, the semi-arid tropics include countries of western and southern Africa plus regions of Brazil and India. It might therefore make sense for these countries to coordinate their national research efforts. Indeed, in this region coordination is found in various commodity networks such as sorghum and millet research networks (the International Sorghum and Millet network "INTSORMIL," the Cereal and Legumes Network in Asia "CLAN," and the West and Central African Sorghum Research Network "WCASRN") and in networks focusing on resource issues such as drought tolerance (Réseau International de Recherche sur la Résistance à la Sécheresse "R3S"). However, not all transnational collaboration is based on similar agroecological conditions. Countries may wish to collaborate for other reasons, some of which will be explored further.

Some authors (e.g., Eicher 1989) cite the experiences of regional research organizations operating during the colonial period in Africa to show regionalization's benefits. Former French and British research institutes specialized in specific commodities such as cotton, coffee and cocoa, oil palm, and rubber. Their application domain extended throughout the area in which production of their mandate commodity was feasible, irrespective of administrative boundaries. For example, in colonial times a single institute led research on oil palm in what is now four different countries: Ivory Coast, Cameroon, Benin, and Togo. This integration was largely facilitated by the unified political framework under which these institutes operated.

Over the past 30 years, several efforts at transnational regionalization have emerged. Networks are the most common mode of regionalization, but a number of truly regional research organizations have sprung up as well. This chapter classifies regionalization efforts in three main categories: topic-based networks, organization-based networks, and regional institutes.

Topic-based networks

The most common form of regionalization of research has been the single-topic network (Plucknet et al. 1990), which generally focuses on a commodity (such as cassava, beans, or groundnuts), a constraint (rust, drought, white fly), a resource (soil management or animal traction), or on a particular research practice (farming systems research). These networks may involve a combination of functions, including information and material exchange, scientific consultation, and collaborative research. They are often based on partnerships among scientists with similar specializations from different countries who collaborate on behalf of their organizations. Topic-based networks tend to be managed from a central point, with minimal interaction among participants. The international research centers of the Consultative Group on International Agricultural Research (CGIAR) have often provided management of topic-based networks in agricultural research.

Organization-based networks

Organizations may choose to cooperate in networks on specific commodities or themes, seeking to perform functions similar to topic-based networks. The principal difference between these networks and topic-based ones is that interactions among organizations are more systematic. Participating organizations also have greater ownership of network activities and generally see the networks as instruments for their own development. Subregional organizations such as CORAF, SACCAR, and the PROCIs in Latin America are of this type. Research remains in the individual member institutions, and a secretariat is established to coordinate activities entrusted to the network.

Regional institutes

In the regional research institute, participating organizations have agreed to devolve a particular research activity to a regional body. Two cases illustrate different evolutions in the creation of a regional institute. The first, WARDA, was set up by West African governments to promote and undertake research on rice. Its original mandate has remained, but has widened since its admission to the CGIAR. The second example, CATIE, in Costa Rica, was initially established by IICA, a regional organization for the Americas. It progressively became independent of IICA, however, and is now governed by its own board and a general assembly composed of ministers of agriculture primarily from Central American countries. In both cases, extensive research infrastructure was built to meet a perceived need for research at the regional level.

Planning regional research

The "subsidiarity" principle provides a starting point for planning regional research. It means that problems are best solved in the subsystem where they arise. Applying subsidiarity to regional research institutes and agendas raises two potential pitfalls. First, participating members may lack sovereignty because of funding constraints. External funding may lure them into regionalization initiatives to solve problems that typically would fall in their own domain. Second, the subsystem must have the authority to effectively address the issues relegated to it. A political oversight body may strengthen the legitimacy of a regional research institute, as long as it doesn't create a bureaucratic or procedural overload. SACCAR and INSAH are examples of regional institutes for which a council of ministers from member countries approves programs of activities.

Regional responsibilities

Organizations need to ask how much responsibility they wish to delegate to a regional entity. Not all research problems should become regional. Only issues that cannot be tackled effectively at the national level, because either the costs are prohibitive for one country or the solution to the problem requires some regional coordination, should become the core of regional programs.

A key dimension in planning regionalization efforts is the extent of integration and shared commitments to be pursued by the different participating organizations. This may be described in four levels. These are, in order of increasing commitment to regional objectives, information exchange, coordination, institutional collaboration, and integration (Plucknet et al. 1992).

Information exchange. This refers to informal, voluntary exchanges of information between individuals through conferences, field trips, newsletters, correspondence, and exchange of genetic materials. Such events, often organized ad hoc by donor agencies, may lead to other, more intensive forms of regionalization.

Coordination. Coordination requires a somewhat greater commitment of participating organizations, which make their own decisions but take into consideration activities and programs of others (Elliott 1994). Examples of coordination at the regional level include the construction of international databases on scientists and current research, organization of regional training programs, and consultations on regional research priorities. Here, savings from eliminating duplication are achieved voluntarily by recognizing and sharing outputs of work executed elsewhere in the region. The cost of coordination is not negligible, however, and may be a limiting factor in establishing effective coordination mechanisms.

Institutional collaboration. Collaborating organizations agree to curtail some activities at the national level and to strengthen certain others in order to avoid unnecessary duplication and to capitalize on each other's comparative advantages. In principle, such collaboration should lead to regional research programs with a division of work dependent on relative strengths and the commonality of problems faced.[1] In practice, few such regional programs have emerged, probably because they require some political endorsement.

Integration. In this, most intensive form of regionalization in agricultural research, research organizations and systems come together to create a new research institute with a regional mandate to conduct research in its own facilities with its own resources. CATIE in Central America is an example of this type of regionalization. INSAH is another example, having a regional mandate to coordinate agricultural research in the region.

In practice regionalization initiatives may traverse a number of the levels described above. For instance, some networks foster information exchange and coordination, while others with more resources may include elements of collaborative research. Integrated regional efforts often include elements from each level.

Implications for planning

Several aspects set regional research programs apart from national programs, leading to special requirements for planning. First is the need to fully understand capacity in the participating national organizations. It is easy to overestimate national technical, financial, and management capacities in the planning stage of a regional effort. Yet scientific leadership may be unavailable, administrative backup may be lacking, and local accounting practices may be inadequate to ensure full accountability. These obstacles could reduce participants' ownership of the regional program and lead them to lack commitment to its goals.

The choice of research topics is also central to the success of a regional research program. As Lele (1998) puts it, there is yet insufficient analysis of the value added in regionalization to guide the choice of research topics. Topics should provide opportunities to capture economies of scale for countries with similar agroecological conditions through spill-in and spill-over effects at acceptable levels of transaction costs. Methodological development in dealing with issues of regional interest, such as natural resource management (NRM), is a likely candidate. Methodological development could produce results that enhance national capacities in conducting NRM research. While the above sets the general conditions for identifying research topics, it is important that each participant gains enough to ensure their continued support to the program.

Managing the internal cohesion of regional research efforts is a third important aspect. Internal cohesion typically depends on the perceived net payoff of the regional effort. It can be threatened by loss of commitment to program objectives or by disagreement on the allocation of resources. Sharing resources is a difficult issue in any form of federated system, including regional organizations. It requires transparent allocation mechanisms, strict budget procedures, and strong accountability based on standard accounting practices. Putting such management systems in place may require some initial assistance, but it is nonetheless essential for the continued viability of regional efforts. Start-up difficulties are alleviated when external donors provide a share of the budget, but they cannot be overlooked if the regional program is meant to become sustainable over time.

Typically regional organizations bring together people of different cultures and disciplinary backgrounds. However, the implicit mode of functioning in one culture is not always transferable to another. Misunderstandings can easily arise. Time for interaction must be allotted to ensure that issues are properly debated and common understanding is reached. If managed appropriately, differences become a source of enrichment as diverse perspectives merge.

Further, planning regional research should allow for a sense of ownership among participating organizations. Participants' ownership of a regional institute is acquired through their intellectual and financial contributions in operational matters and in the design of research programs. Ownership also requires that benefits accrued from working together are truly shared. Gains from regional efforts lie principally in access to results that one country or organization could not achieve alone at a reasonable cost. It therefore stands to reason that such access must be ensured from the outset of the collaborative effort. This can be a thorny issue in sensitive research areas.

Regional efforts may come under fire at the national level. The most likely attack is that the project is not yielding the promised benefits. This problem can be avoided by establishing at the outset a process for evaluating success or failure according to preset goals. The individuals most involved in regional programs may also come under fire if complete transparency in choice of scientists and projects is not ensured. National research organizations should ensure their scientists are active in the governance mechanisms put in place to oversee the regional program. Such mechanisms cannot be left in the hands of management or a few entrepreneurial individuals alone, because then the regional initiative may be discredited and, as a result, its benefits left unutilized within the country.

Tight budgets and communication problems often play against scientists assigned to remote areas. Yet if they are to take issues of local context seriously, regional initiatives should take care to involve scientists working in isolated field stations. It is easy for "regionalization" to become a means for researchers from capital cities to get together: nice for travel experiences but less effective for addressing constraints facing their countries' agricultural sectors. Failure to sustain a focus on locally relevant issues could lead to further discontent at the national level, thus reducing support for regional initiatives.

Dependable, sustained financing is at the heart of successful regional research institutes. Such financing may combine core contributions from the participating countries or a higher level regional institute with program or project funding from donors. Core contributions should allow the regional initiative to pursue a strategy that reflects the interests of its main partners. Inability to meet basic operating costs may put the members at the mercy of donor schedules and agendas and raise questions regarding national political commitment, leadership, and long-term viability (Eponou 1998). Regional bodies should be lean so that their basic operations can be financed within the means of their main partners. Any expansion of the organizational structure should be conditioned by the ability to cover at least basic operating costs.

Finally, partners in a regional initiative must be sufficiently compatible in terms of size, interests, and objectives. Partners working on common crops or farming systems stand a better chance of collaborating successfully. Regional bodies must avoid rapid growth in responsibilities and programs. Many have suffered from mission creep, that is, they have attempted to expand their mission beyond what was initially agreed. The existence of a political oversight committee lends legitimacy and national organizations' participation in regional initiatives enables it to make binding decisions and reduces the chance of mission creep.

Relevance for agricultural research

Improving the performance of research organizations and programs is a prime reason for both research organizations and their funding agencies to foster and adopt regional modes of collaboration. Regionalization may increase the effectiveness of research programs through the development of appropriate research tools. For instance, regional collaboration is useful for honing tools and methods of interaction with farmers or developing research methods for natural resource management. Such methodological development, though not of immediate concern to users, will eventually impact the technologies proposed to use them.

The direct impact of regionalization on the delivery of new technologies to farmers depends on the ability to focus a critical mass of scientists mobilized from different countries on priority issues that are relevant to users in the various countries. Only if the problems facing farmers in different countries are truly comparable will regional programs be able to impact farmers directly. Where problems are different but require a similar knowledge base to resolve, regional programs may support national or more location-specific technology generation initiatives. In such cases, regional research institutes will need to collaborate with national research systems where these have established effective contacts with users.

It is in improving *efficiency* of research that regionalization can make its greatest contribution. Information sharing is often the first step in this respect. Knowledge exchange is vital for countries whose resource base is too narrow to allow them to produce knowledge and technologies in all domains required. These countries may tap information sources in search of results of significance to their users, rather than trying to develop new technologies single handedly. Sharing knowledge is a way to access information produced elsewhere for similar ecological conditions.

Though the benefits of information sharing are obvious and significant, there are few successful cases with a lasting track record. The reason is that the costs of setting up and maintaining such programs are generally underestimated. Good journals need peer review, editorial committees, and editors. Databases on current research can be built as a one-time effort, which is costly enough, but their maintenance is even more demanding. Websites do not diminish the need for dedicated, motivated people to maintain them.

Collaborative research on topics of shared interest is a more advanced form of regionalization. The most common form is the commodity research network. It is often designed as a tool for coordinating work over a given region. It also is a mechanism to provide technical assistance, for example, in preparing better project proposals, exchanging results on a common theme, and writing and disseminating research reports. Efficiency gains stem from the elimination of duplication: because of improved coordination, research results are shared among countries and need not be repeated in each one; because of better research, experiments are more conclusive and need not be repeated over time. These networks are also attractive to donors who want to fund research but see little merit in grants to maintain organizations.

Some research topics require a regional approach. Developing adequate defenses against pest and diseases calls for a regional offensive. For example, the fight against striga in West Africa cannot be tackled by one country working in isolation. Joint research on new topics that require significant initial efforts is yet another form of regional collaboration. Start-up investments for research on an emerging topic are often substantial, requiring a wealth of human expertise and capital. Furthermore, such research may become effective only after an extended period of trial and error. By sharing expenses and trying out different ideas across countries, the costs of learning are reduced.

Political reasons

By bundling their voices in regional and subregional bodies such as CORAF, ASARECA, APAARI, and the PROCIs, countries obtain more influence in international fora. For example, regional institutes now wield considerable authority in managing the global research network and setting the agenda for the CGIAR research group.

Economic and political integration in the European Union started with the integration of specific industries such as steel and coal. Agricultural research has never been the single spark for integration. But it has been instrumental in some efforts towards greater regional integration, such as SACCAR in Southern Africa.

Donor interests

The pressure to innovate in research funding through regionalization comes from the current scarcity of aid funds and competing demands for funds from sectors that can show rates of return equally impressive to that of agricultural research. Donor agencies have turned to regional networks as a way to get the most for their limited means. Regional networks allow donors to reach a large number of scientists in several countries, giving higher visibility to their support. Moreover, transfer of coordination responsibility to regional organizations can reduce donors' costs in managing many small research grants, which can be very demanding in terms of administration time within the donor agency. With donor funding, regional networks gain greater ownership in network design and management. Using regional mechanisms also enables donors to bypass national organizations and yet provide substantial financial and technical support to scientists.

Regionalization is multifaceted, and assessing its usefulness for agricultural research systems depends on the subject matter being regionalized and on the funding mechanism used. The net benefit of regionalization to an organization depends on how much it has to pay, not only in monetary terms, to reap benefits. Unless such costs are considered, it is easy to overestimate regionalization's net benefits and sustainability.

Fostering greater interaction and communication among organizations and scientists is one of the prime benefits of regionalization. Journals, conferences, and symposia have been providing such interaction over the years and, in as much as regional mechanisms foster such exchange, they are beneficial to scientists and their organizations. Linked to this free flow of ideas is the potential for capitalizing on the diversity of views and experience in a region, to explore a new field of inquiry, and to build up intellectual capacity. Another significant benefit is the possibility of accessing costly technologies through regional efforts. The Centre d'Etude Régional pour l'Amélioration de l'Adaptation à la Sécheresse (CERAAS) is a base center hosted by ISRA and sponsored by CORAF. CERAAS provides scientists in the region opportunities to use sophisticated equipment in a well organized laboratory technically supported by an advanced research institute, in this case CIRAD.

On the downside, regionalization may carry heavy transaction costs. Policy-making at the ministry level, planning at lower levels, evaluation mechanisms, and follow-up to evaluations require enormous numbers of meetings that entail heavy communications and travel costs in addition to the time investment. The introduction of Internet-related technologies will enable virtual meetings and may reduce the number of face-to-face encounters, but phone, fax, and e-mail costs will remain high in the foreseeable future.

Two final issues influence the relevance of regional research to agricultural research systems. First, the balance between ad hoc and long-term collaboration must be carefully assessed. Many regional research efforts owe their existence to donors and international organizations who wish to fund research on issues they consider important. While such ad hoc networks may have a place and even be an acceptable step towards regionalization, their start-up costs might be inordinately high. This would argue in favor of creating regional platforms to support ad hoc activities, thus separating the regional mechanism from the activities it sponsors.

The range of partners involved in regional research is also becoming a strategic issue. National-level, public-sector agricultural research organizations have often been the initial partners in many regional and subregional initiatives. With a growing number of entities, public or private, active in agricultural research in each country, the trend is a widening set of partners, in particular, towards inclusion of universities and nongovernmental organizations (NGOs). The broadening of the research community may pose some organizational problems however. If all potential national partners are to be included, regional organizations could become unwieldy and transaction costs would mount even higher. Some form of national representation in regional bodies must then be contemplated. Including advanced research institutes in regional organizations could be beneficial if regional programs attempt to do advanced research, upstream of national programs. Mechanisms can be devised to ensure that the more advanced partners do not become overly dominant in the regional programs but, rather, remain an indispensable source of know-how and technical assistance.

Examples

PROCISUR, as the other PROCIs in Latin America, was initially created as a flexible mechanism for cooperative research and information exchange. Each participating country was to retain its own management responsibility and programming independence. Its original structure included research programs on a number of commodities – cereals, oil seeds, and cattle – operating under an integrated secretariat provided by IICA. PROCISUR has now shifted to a thematic focus, including subprograms on biotechnology, natural resource management, agroindustry, genetic resources, and institutional development.

Overall priority setting, resource allocation, and supervision of activities are responsibilities of the directors of the participating national research organizations, who meet at regular intervals, usually twice a year. PROCISUR actively mobilizes domestic and international resources to conduct research and technology transfer of mutual interest to participating countries. Eighty percent of its core resources are presently contributed by its members. Its regional program was evaluated in 1992, at which time internal rates of return were calculated at well above 100 percent.

In 1999 PROCISUR was widening its range of partners to include organizations from the private sector and universities. Its leaders expect this to increase the relevance and effectiveness of its research projects, thereby further strengthening its role in the agricultural technology system of the Southern Cone.

Institut du Sahel

The Institut du Sahel (INSAH) was created in 1977 by the Council of Ministers of the Sahelian countries in response to the drought that hit the Sahel region from 1968 to 1973. Though each country had its own public-sector agricultural research organization, it was felt at the time that none properly focused on themes of food self-sufficiency and desertification control. INSAH was meant to help build up the member states' research capacities and to coordinate research activities to avoid duplication in key areas. Its mission did not include research per se but coordination, as well as information sharing within the region, technology transfer, and training of technical staff (Jallow 1992).

INSAH is one of the specialized institutions of CILSS, the French acronym for the "Permanent Interstate Committee for Drought Control in the Sahel." Participating Sahelian governments contribute between five and 10 percent of INSAH's budget. More than two-thirds of the budget is channeled directly to support field research in national organizations participating in collaborative research programs. This funding is often matched by national funding. Such programs are planned in collaboration with organizations in participating countries.

INSAH has created a bibliographic database RESADOC on research in the Sahel and also provides small travel grants to scientists to visit research organizations in the region. Training programs include those on natural resource management, socioeconomics, technology transfer, and management training. Recent research programs focused on socioeconomic research: food security and environmental analysis.

Donors supporting CILSS are organized in the Club du Sahel, which brings greater cohesion to funding. This group of donors plays an active role in guiding CILSS and indirectly INSAH through the research they undertake on their own and by the role they play in governance. Regular program and management reviews are funded by donors and are conducted at their request.

References

Eicher, Carl. 1989. Sustainable Institutions for African Agricultural Development. ISNAR Working Paper No. 19. The Hague: International Service for National Agricultural Research.

Elliott, H. 1994. Coordination of research: Issues, experience and lessons from outside Africa. Paper presented at SPAAR workshop on regionalization in agricultural research in West and Central Africa. Discussion Paper No. 94-6. The Hague: International Service for National Agricultural Research/Special Program for African Agricultural Research.

Eponou, T. 1998. Financing research through regional cooperation. In *Financing Agricultural Research: A Sourcebook*, edited by S. R. Tabor, W. Janssen, and H. Bruneau. The Hague: International Service for National Agricultural Research.

Gijsbers, G. and R. Contant. 1996. Regionalization of Agricultural Research: Selected Issues. Briefing Paper No. 28. The Hague: International Service for National Agricultural Research.

Jallow, A. T. 1992. Regional Collaboration in Agricultural Research: The Experience of the Institut du Sahel. The Hague: International Service for National Agricultural Research.

Lele, Uma. 1998. Building regional cooperation from the bottom-up and top down: The case of the Southern Cone countries. In *Guidelines for Designing New Organization and Funding Ways for Agricultural and Agroindustrial Innovation Systems in the Southern Cone*. Montevideo: Inter-American Institute for Cooperation on Agriculture.

Plucknet, Donald L., N. J. H. Smith, and S. Ozgediz. 1990. International Agricultural Research: A Database of Networks. CGIAR Study Paper No. 26. Washington, D. C.: World Bank.

Plucknet, Donald L., S. Ozgediz, and N. J. H. Smith. 1992. Assessing current and potential IARC/NARS Networks: A focus on institutional impact. In *Assessing the Impact of International Agricultural Research for Sustainable Development*. Proceedings from a symposium at Cornell University. Cornell: Cornell International Institute for Food, Agriculture and Development.

Chapter 3
Integrating Natural Resource Management in Agricultural Research Planning

Gerdien W. Meijerink

National agricultural research organizations, which have for a long time concentrated on productivity enhancement, are shifting their focus more towards resource management concerns. Although these organizations are experienced in planning commodity and disciplinary-oriented programs, they face new challenges in applying their traditional planning methods to research on natural resource management (NRM). Incorporating NRM into public planning cycles and dealing with NRM's characteristic uncertainty, long time horizons, and multiple stakeholders all present new planning dilemmas. Moreover, because these issues are intertwined, they are difficult to integrate systematically into the planning of a broader agricultural research agenda. This chapter discusses some recent trends in NRM research and suggests how agricultural research organizations might effectively incorporate NRM into research planning.

What is natural resource management-oriented research?

Throughout the world, increasing attention is being devoted to issues of sustainable natural resource management (NRM). Agricultural research organizations that have long focused on productivity enhancement, are now shifting their attention towards resource management concerns. Although many have experience in planning commodity and disciplinary research programs, they face new challenges in using traditional planning methods for NRM (see Crosson and Anderson 1993). NRM invokes complex issues. NRM-oriented agricultural research focuses not only on improving agricultural productivity and profitability, but also on making sustainable and efficient use of the natural resource base. Developing countries especially have a lot to gain by making optimal use of natural resources. To do so, however, the "management" aspect of NRM is critical. Recent years have seen a shift in perspective, from a technical (i.e., biological, ecological) to a more social (cultural, economic, legal) view of NRM. This has implications for planning NRM-oriented agricultural research. Increasingly, integrating NRM concerns into agricultural research is affecting agricultural research planning processes.

Agricultural production activities affect natural resources in numerous ways. At the same time, changes in the environment and changes in the management of natural resources impact agricultural production. Agricultural production and NRM are therefore intertwined and cannot be treated as separate issues in agricultural planning. Although in effect all agricultural research agendas should fully integrate NRM concerns, there remains a distinction between commodity-oriented research and NRM-oriented research. Several dimensions make NRM-oriented agricultural research more complex and difficult to tackle than commodity-oriented research. NRM-oriented research often deals with long time horizons, issues of a public nature, manifold stakeholders, and, usually, more uncertainty than commodity research programs. However, not all NRM-oriented agricultural research scores high on all these points; they are merely general characteristics of NRM. Figure 1 characterizes NRM research according to the "PUSH" scheme. The higher the research scores on the four axes in the figure, the more difficult it is to tackle. Some NRM-oriented research, such as that on integrated pest management (IPM), scores relatively low on the four axes, thus approaching the difficulty level of commodity research.

Public goods

Many environmental goods and services are public in nature. As with many other public goods (e.g., health), they are linked with market failure – that is, there is no market in which they can be traded and where a price is determined. This often calls for government intervention, and although government may not always be the most appropriate body to deal with market failure (besides market failure there is also government or policy failure), governments are instrumental in creating an environment in which market failure can be resolved. There are several market failures in NRM. One of these is the existence of "externalities," that is, when the effects of a local activity are felt by third parties outside that locality. For instance, local land use, such as that which causes deforestation, often has wider impacts, such as disappearing watersheds or biodiversity. Another example of market failure is the difficulty of putting a monetary value on environmental goods and services.

All these issues have important consequences for NRM-oriented agricultural research. Private goods and services are easier to deal with because they can be costed, have a clear target group, and (intellectual) property can be safeguarded. The private sector is thus likely to dominate research in these areas. NRM is not as easy to sell as, for example, seeds. The fact that NRM is of public relevance but is unlikely to pay off commercially is a strong argument for it to be financed by public funds (see also Byron and Turnbull 1997). In reality, however, na-

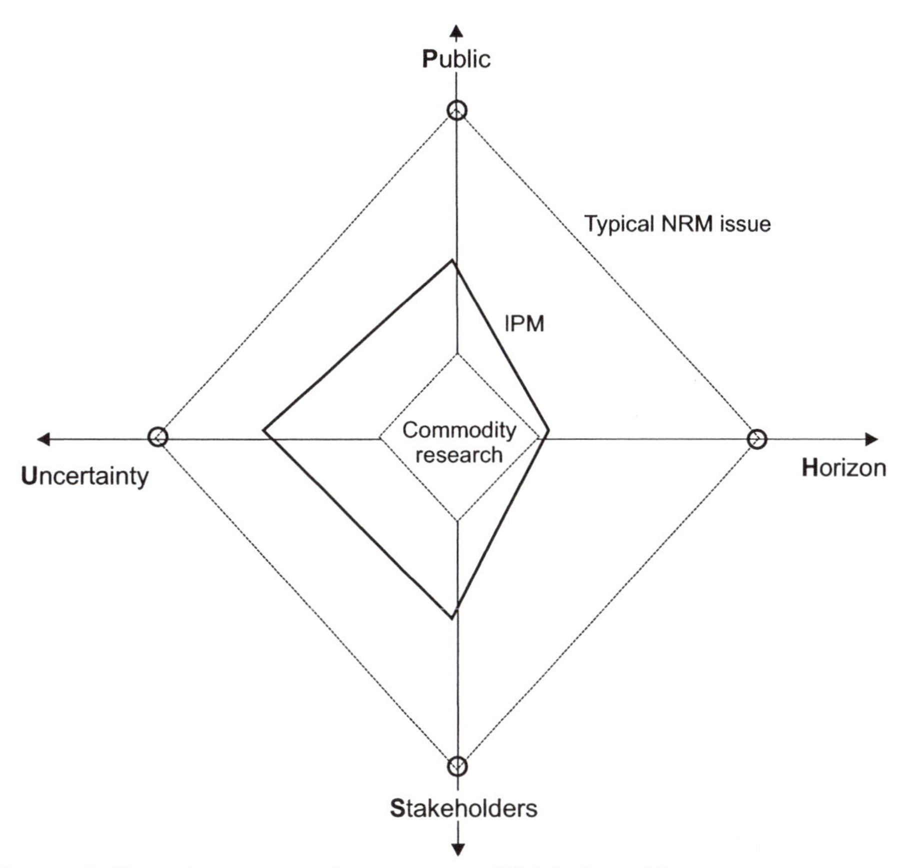

Figure 1. Four dimensions characterizing NRM: the public-uncertainty-stakeholders-horizon (PUSH) model

tional governments are becoming less involved in (agricultural) research, and in some countries funds for public bodies such as national agricultural research organizations have diminished over the past decade.

In NRM-oriented agricultural research, outputs are not only beneficial to the farmer but also to society at large. This implies that policy analysis and design are more important in NRM work than in production-oriented agricultural research. Whereas traditionally, agricultural research thought of farmers as its most important client group, NRM research also emphasizes governments and policymakers as client groups. However, links between leaders of agricultural research programs and policymakers are often weak and in need of strengthening (Tabor et al. 1998). Involving policymakers in the planning process could be an important step forward.

Uncertainty

NRM research is associated with risk and uncertainty. Due to its often long time horizons, complexity of agroecological systems, lack of relevant information and knowledge, and short empirical track record, it is difficult to predict what effect changes in use of land and natural resources will have on the environment in the future. This is especially evident in the field of biodiversity: the Earth harbors between 10 million and 80 million species, of which only about 1.4 million have been identified (Ryan 1992). Our understanding of the complex relationships between species and what effect extinction of one species might have on others and on the ecosystem as a whole is as yet incomplete. We do know that the consequences of irreversible loss are great, "as genetic variability is lost, … the species as a whole becomes more vulnerable to other factors, more susceptible to problems of inbreeding, and less adaptable to environmental change" (PBRP 1992: 18). But we do not know when losses are irreversible, what the thresholds are, or what NRM decisions are "safe."

Uncertainties due to the complexity of systems and lack of information play a role in many NRM areas and have significant implications for technology design. Users need to be able to apply and adapt technological options in the uncertain and variable circumstances they face. There are various answers to these uncertainty problems. A first is research to describe and understand resource degradation. Second are researchers' attempts to err on the safe side; when they are unsure of the consequences and reversibility of degradation, only a very limited extent of degradation is allowed for new technologies. Third, new technologies are not "finalized," then diffused. Rather, they are fine-tuned and modified alongside users during the research process. "Adaptive management," a form of management that aims to take risk and uncertainty into account, is discussed later in this paper.

Multiple stakeholders, users, and sources of innovation

The environment performs multiple functions (related to a range of goods and services). This implies that a multitude of stakeholders make direct or indirect use of various functions. Some of the different uses are conflicting. For instance, water that is used for irrigation or sewage cannot be consumed. Use of a forest for timber may conflict with maintenance of its biodiversity or its watershed function. Conflict management is therefore a major challenge in NRM. In this respect it is important to note that use of the different functions is the outcome of rational decisions by users who are influenced by their context (including administrative, policy, natural, and socioeconomic dimensions). Understanding the context is crucial for getting a grip on causes of resource degradation. Too often, population growth and poverty are seen as the causes of environmental degradation. However, these factors alone are rarely the whole cause of degradation, and such simplicity of analysis undermines effective conflict resolution. Creating an "enabling environment" – a context that allows users to make rational decisions about the sustainable use of natural resources – is increasingly recognized as instrumental for resource conservation.

Multiplicity of stakeholders also means a multiplicity of sources of innovation (see Biggs 1990), which is increasingly common in NRM. Especially under stress and conflict, users are developing more efficient practices and innovating in organizations and institutions such as common property (e.g., joint forest management).

All this has consequences for NRM-oriented agricultural research. One implication is the reduced importance of technical solutions. This goes hand in hand with an increased scope for management and social solutions such as institutions (in the sense of a set of rules) and platforms where stakeholders meet and reach agreements. Another implication is that stakeholders' concerns and priorities should be incorporated into the NRM-oriented agricultural research agenda. Although participatory approaches are entering the mainstream, early involvement of (small-scale) farmers in agricultural research is not yet common practice.

Long time horizon

An important element in NRM-oriented agricultural research is concern for future generations. As stated by the Brundtland Commission (1987, p.43), "sustainable development is development that meets the needs of the present without compromising the ability of future generations, to meet their own needs." Use of natural resources has implications for future generations and it is important to know what these effects will be. Seldom can our actions be righted in the short term. Destroyed tropical rain forests or natural flood lands, for example, require centuries to be restored to what they once were: complex and fragile ecosystems with a wealth of biodiversity and enormous biomass. Improving degraded soils is another slow process whereby not only fertility must be restored but also organic structure.

Linked to long time horizons is the aspect of risk. Some degradation processes lead to irreversible effects that cannot be righted, such as extinction of a species, desertification of agricultural lands, drying up of lakes (such as the Aral lake in the former USSR), or depleted fish stocks. Such damage has enormous income consequences for resource users. The agricultural sector, as a major user of natural resources thus has a lot to gain from efficient and sustainable use of natural resources. NARS have an important role to play in generating (agricultural) technologies and management practices that make better use of natural resources.

NRM as an integrated concept

Although we have distinguished four dimensions, it is clear that they are intricately linked. NRM is a complex concept, making it difficult to isolate different natural resource issues. An ecosystem, such as a forest, is more than the sum of its parts. A change in one of the parts will change the whole system (although, of course, not all effects are substantial or significant). Changes in natural resources or the environment also affect human beings in several different ways. This makes planning NRM-oriented agricultural research a complicated task. After all, a solution in one area may cause problems in another. Expansion of irrigation, for instance, may increase mosquito populations and malaria incidence. NRM-oriented research must find ways to deal with complexity. A holistic approach is often necessary, but difficult to implement. The effects of NRM on disciplines other than agriculture is often overlooked by agricultural research organizations. Addressing such complex issues, however, calls for concerted action involving an array of research entities.

Integrating NRM concerns into planning

Agricultural research plans reveal an increasing focus on NRM since the early 1990s. However, even in the 1990s, efforts to incorporate NRM into agricultural research were limited to inquiry into the effects of agricultural production on the environment (e.g., externalities), and NRM issues often remained an isolated sideline in the plans or were treated as *nota bene*. There are some exceptions where particular effort has been made to integrate agriculture and NRM (examples appear later in this section). Integrating NRM concerns into agricultural planning means taking into account the issues discussed above: long time horizons, public concerns, multiple stakeholders, and uncertainty. This is no easy task, but some directives and suggestions can be mentioned.

Mix of research

As was determined in the PUSH model (figure 1), NRM research often implies focusing on public issues, taking uncertainty into account, considering multiple stakeholders, and incorporating a long time horizon. These characteristics imply that NRM research is a typical activity for the public sector. Yet NRM may lack immediate results or benefits that could cement political support for it. Still, not all NRM research scores high in all the "PUSH" dimensions. For example, IPM offers opportunities to arrive at measurable benefits for a clear client group in a short period of time. To ensure sufficient support for NRM-based research it makes sense to combine NRM research that scores high in the PUSH model with research that scores low in the model. However, if this type of research lends itself to private funding and is already undertaken effectively and in a responsible manner by the private sector (e.g., agrochemical companies may invest in IPM), the public sector should not compete. There are plenty of other NRM research needs among poor people that the private sector will not take on.

Information and data needs

For planning, research organizations need the right information about physical trends (e.g., biological, soil, climate) and social trends (e.g., understanding how resource users make decisions, recognition of innovations in NRM). There are different strategies for obtaining information and knowledge. In modeling approaches, mathematical constructions using data are applied to generate information, for example, by developing different future scenarios. Using participatory methods, information is gleaned through "dialogues" between different stakeholders, as in participatory rural appraisal techniques. But as Alsop and Farrington (1998) state, getting information from multiple stakeholders can be unmanageable and time consuming. They advocate a network of "nested" monitoring systems. Recently, there has been a trend towards integration of methods (e.g., combining modeling with participatory approaches). As data gathering is often time consuming and costly, establishment of minimum data requirements is important and the purpose of collecting data should be clear.

Participation of stakeholders

Planning for NRM-oriented agricultural research requires participation of a broad array of stakeholders: farmers, national and local policymakers, private sector, nongovernmental organizations, and representatives from ministries and research organizations other than for agriculture. They will put forward specific objectives and priorities, as well as contribute new insights and information. Platforms can be tools for participation. However, new platforms are not always necessary; use can be made of existing platforms such as national environmental action plans or planning fora at a provincial levels.

Tackling complex issues

NRM's complex and uncertain character calls for an integrated, holistic approach. However, in practice this is difficult to achieve and extremely difficult to manage. Breaking complex issues into smaller, manageable problems may be useful but also increases the risk of reductionism. An alternative approach is to use adaptive management (see Holling 1978; Walters 1986; Lee 1993), which combines management decision making with learning and improving information. This latter approach is especially useful in complex situations where uncertainty and a paucity of information prevail.

Adaptive management has been applied to a range of issues including fisheries and forest and water management since its development in the 1970s. Complex problems are divided into a set of "experiments" on possible management decisions. By divining and learning from the outcomes of these experiments, management policies and practices are improved. Although this technique has been applied mainly in the North, it is also applicable in the South (see Gunderson et al. 1995). Experience with it has been gained in joint forest management in India (Pfoffenberger and McGean 1996).

Relevance for agricultural research organizations

As the world's population grows, pressure on sometimes scarce natural resources increases. Agricultural research organizations have an important role to play in relieving this pressure by coming up with feasible solutions: technologies or management options that make efficient use of the natural resource base. Progress has been made particularly in managing forest, water, and soils, as well as in integrated pest management and biodiversity maintenance.

Forest management

Forestry is a traditional NRM area within agricultural research. However, the focus of forestry research has changed considerably over the past decade, shifting from improving timber production to acknowledging that a forest performs many more functions beyond its production role. Several forest functions are related to agriculture, such as biodiversity maintenance, watershed management, and supply of food, fodder, and other nontimber forest products. Current research focuses on sustainable management of these different functions rather than on timber production alone (see also Byron and Turnbull 1997).

Water management

As countries' populations grow, fresh water is becoming scarcer and water resource management is becoming increasingly important (UNEP 1999). Fresh water comes from an essentially closed system. The amount of water available is therefore more or less fixed. Research has thus shifted focus from water-supply management to water-demand management (Seckler 1996). Conflicts over water are expected to intensify as water consumption increases. Users and uses of water are multiple. Agriculture uses about 70 percent of available water. Of this, irrigation is a main use, of which 17 percent is in developing countries (FAO 1995).

Soil (nutrient) management

Soil is one of agriculture's main natural resource bases. As a country's population increases, arable, fertile land becomes scarcer, resulting in intensification of land use and shorter fallow periods. With elimination of fertilizer subsidies in many places and limited availability of organic fertilizers (e.g., manure), nutrient replenishment and maintenance of organic structure become problematic. More emphasis is therefore being put on integrated nutrient management (in which livestock, agroforestry, household waste disposal, and crop production are integrated) and on soil and water conservation. Besides soil fertility maintenance, soil erosion is a major problem in many countries. Erosion not only diminishes fertility, but in the case of erosion through water, also causes problems downstream by clogging waterways.

Integrated pest management

Pesticides were long seen as an important means of increasing productivity. They were therefore promoted by agricultural research organizations. However, evidence continues to surface on the negative impacts of pesticides on the environment and on human health. This has led many countries to ban certain pesticides, and increasing attention is now being given to safe, environmentally friendly alternatives. IPM is one. IPM integrates a number of pest-control techniques to discourage the development of pest populations and keep pesticides limited to levels that are economically justified and safe for people and the environment (FAO 1998).

Biodiversity

The importance of biodiversity is now recognized and concern is swelling over the rapid loss of species. Although there is still a paucity of knowledge about the number of species and exact rates of extinction, it is clear that biodiversity is being threatened by the disappearance of high-biodiversity habitats. Expansion of agriculture has contributed to the loss of habitats (e.g., through deforestation of tropical rainforests for agriculture). For agricultural research purposes, scientists increasingly appreciate the importance of biodiversity for research using techniques from advanced cell and molecular biology, often grouped as "biotechnology" (see Falconi, this volume). Biodiversity in the sense of a pool of genetic resources is crucial for this type of research, and new biotechnology techniques have strengthened efforts to characterize, manage, and use biodiversity (i.e., genetic resources). Genetically modified crops used in agriculture are also seen as a threat to biodiversity, however. The fear is that genetically modified genes (transgenes) could escape from the modified crops into the nonagricultural environment and create, for example, "super weeds" that displace native vegetation.

Examples

Several NARS have undertaken planning incorporating an explicit NRM focus. Such planning usually follows a procedure similar to that used in purely commodity-oriented research planning. However, the four issues that feature in the PUSH model play a more pronounced role. The examples that follow are from ISNAR's experiences and provide an overview of some of the various planning exercises.

Benin: planning and priority setting for regional research

Regional research programs, because of their intrinsic holistic perspective on resource use, typically take NRM concerns into account. Planning of Benin's research program for its southern region was no exception. Objectives for the regional research program were threefold: intensify agricultural production, generate employment in agriculture-related activities, and reduce degradation of the natural resource base. The planning method was based on the program formulation steps described by Collion and Kissi (1994, see also Collion, this volume). The holistic perspective, however, demands careful management of program resources and careful formulation of regional programs to avoid problems of management and focus (see Janssen and Kissi 1997).

Kenya: priority setting for a soil fertility and plant nutrition program

The Kenya exercise aimed to provide guidelines for future resource allocation in the soil fertility and plant nutrition research program of the Kenya Agricultural Research Institute (KARI). Quantifying factor-based research impacts was a crucial but difficult step in planning because of the uncertainties involved in factor-based technologies, such as externalities and the long time horizon. To account for these, five steps were followed: (i) compiling a detailed information base on the mandate area, (ii) identifying research target zones and research themes, (iii) specifying potentials for technology generation and adoption, (iv) identifying and quantifying benefits accruing from research themes and ranking research alternatives, and (v) establishing priorities along with program stakeholders. A simplified approach to evaluating NRM-oriented research was developed and found to add insight into the likely distribution of economic benefits across the program's research themes and target zones (see Kilamba et al. 1998).

West-Africa and Peru: choosing tree species for genetic improvement

This exercise, undertaken jointly by the International Centre for Research in Agroforestry (ICRAF), ISNAR, and national-level agricultural research organizations in the humid and semi-arid lowlands of Africa and Peru during 1993–95 aimed to provide reasonably objective and systematic procedures for dealing with a broad range of issues. The end goal was to find the best possible set of research activities in the face of the information constraints common to NRM-oriented priority-setting exercises. The exercise involved seven (flexible) steps: (i) team building and planning, (ii) assessment of client needs, (iii) assessment of species used by clients, (iv) ranking of products, (v) identification of a limited number of priority species, (vi) valuation and ranking of priority species, and (vii) final choice (see Franzel et al. 1996). In each step the number of species considered was reduced.

References

Biggs, S. D. 1990. A multiple source of innovation model of agricultural research and technology promotion. *World Development*, 18 (11): 1481–1499.

Byron, N. and J. Turnbull. 1997. New Arrangements for Forest Science to Meet the Needs of Sustainable Forest Management. Bogor, Indonesia: Center for International Forestry Research.

Collion, M. -H. and A. Kissi. 1994. Guide d'Elaboration de Programmes et d'Etablissement de Priorités. Research Management Guidelines No. 2. The Hague: International Service for National Agricultural Research.

Conway, G. 1985. Agroecosystem analysis. *Agr. Admin.*, 20: 1–55.

Crosson, P. and J. R. Anderson. 1993. Concerns for Sustainability: Integration of Natural Resource and Environmental Issues in the Research Agendas of NARS. Research Report No. 4. The Hague: International Service for National Agricultural Research.

FAO. 1995. FAOSTAT database (online or CD-Rom). Rome: Food and Agriculture Organization of the United Nations.

FAO. 1998. FAOSTAT database (online or CD-Rom). Rome: Food and Agriculture Organization of the United Nations.

Franzel, S., H. Jaenicke, and W Janssen. 1996. Choosing the Right Trees: Setting Priorities for Multipurpose Tree Improvement. ISNAR Research Report No. 8. The Hague: International Service for National Agricultural Research.

Gunderson, L. H., C. S. Holling, and S. S. Light (eds). 1995. From Barriers and Bridges to the Renewal of Ecosystems and Institutions. New York: Columbia Press.

Holling, C. S. 1978. Adaptive Environmental Assessment and Management. London: John Wiley & Sons.

Janssen, W. and A. Kissi. 1997. Planning and Priority Setting for Regional Research: A Practical Approach to Combine Natural Resource Mangagement and Productivity Concerns. Research Management Guidelines No. 4. The Hague: International Service for National Agricultural Research.

Kilamba, D., S. Nandwa, and S. W. Omamo. 1998. Priority setting in a production-factor research program. In *Agricultural Research Priority Setting: Information Investments for Improved Use of Resources* edited by B. Mills. The Hague: International Service for National Agricultural Research.

Lee, K. N. 1993. Compass and Gyroscope: Integrating Science and Politics for the Environment. Washington, D.C.: Island Press.

PBRP. 1992. Conserving Biodiversity: A Research Agenda for Development Agencies. Report of a Panel of the Board on Science and Technology for International Development. Washington, D.C.: National Research Council, National Academy Press.

Ryan, J. C. 1992. Conserving biological diversity, In *State of the World* by L. Brown et al. New York: Norton.

Seckler, D. 1996. The new era of water resources management: from "dry" to "wet" water savings. Research Report No. 1. Colombo: International Irrigation Management Institute (IIMI) (now "IWMI", the International Water Management Institute).

Tabor, S. and D. Faber. 1998. Closing the Loop: From Research on Natural Resources to Policy Change. Policy Management Report No. 8. The Hague and Maastricht: International Service for National Agricultural Research and European Centre for Development Policy Management.

UNEP. 1999. Global Environment Outlook. Nairobi: United Nations Environment Programme.

Walters, C. J. 1986. Adaptive Management of Renewable Resources. New York: Macmillan.

WCED (World Commission on Environment and Development). 1987. Our Common Future. Oxford: Oxford University Press.

Chapter 4
New Technologies and Planning

Cesar A. Falconi

Biotechnology and information technology are changing the way in which agricultural research is undertaken. They allow scientists to address problems that cannot be solved using traditional research tools. Biotechnology and information technology planning, however, requires special attention because of new technologies' particular characteristics. The performance of new technologies is often hard to predict and may lead to previously unheard of problems, such as public acceptance. Changes in the structure of research organizations may be required, for example, to absorb the high capital costs of new technologies. Timing of investments in new technologies is also important, to avoid being locked into expensive technologies that quickly become obsolete. Legal frameworks may need adaptation and staff with certain skills may need to be contracted or trained. In short, integrating the new technologies of the moment and of the future is an important consideration in planning.

What are new technologies?

During the last two decades major technological breakthroughs have occurred that are shaping or will shape the technology strategies of research organizations. The technological base of agriculture and agricultural research are changing from biological to molecular disciplines, from analogue to digital engineering, and from process management under "controlled conditions" to simulation of reality. Biotechnology and information technology are two key components of these changes. These two areas of innovation are changing or will change the shape of agricultural research for many years to come (see table 1).

Biotechnology

According to Cohen (1994), "biotechnology" includes any technique that uses living organisms or substances from organisms to make or modify a product, to improve plants or animals, or to develop microorganisms for specific uses. In agriculture, biotechnology has many traditional applications, such as composting, vaccines to control animal disease, and cheese and winemaking. New are the "modern" techniques and applications in cellular and molecular biology that were derived over the 1980s and 1990s.

Many biotechnology applications are an extension of traditional plant and animal breeding techniques, and they often complement rather than replace long-established methods. The traditional methods, however, are limited to species that are sexually compatible. Biotechnology can expand the range of traits beyond those found in a sexually compatible species. Twenty-five years ago it was unthinkable for plant breeders to transfer into rice plants genes from tomatoes or beans – much less from bacteria. Now with recombinant DNA techniques, genetic transformation of this kind has been successfully applied in a range of agricultural crops. For example, transgenic tomato, tobacco, cotton, and soybean have been developed with pest resistance derived from a group of toxin-producing genes, the so-called *Bacillus thuringiensis* or *Bt* genes from bacterial DNA.

Development of in-vitro tissue and cell-culture techniques has occurred in parallel with advances in molecular biology and genetic engineering. In-vitro techniques make it possible to regenerate a whole plant from a small piece of tissue, and even from a single cell, by growing it in a suitable medium. In research on plants, tissue-culture techniques can be of great value for achieving rapid multiplication of a desirable genotype. Promising biotechnologies for livestock include in-vitro fertilization and embryo transfer.

Table 1. Transformation of Technologies in Agriculture

	Biotechnology	**Information Technology**
Traditional technologies	Fermentation using enzymes and microorganisms, traditional breeding techniques (key science: biology)	Data stored and transmitted in analogue form using electricity and electronics (key science: electrophysics)
Emerging technologies	Genetic engineering; genetic markers; genetic diagnostic, tissue-culture, and microbiological techniques (key science: molecular biology)	Data manipulated and transmitted in digital form using microelectronics, optronics, and associated software (key sciences: physics, computer science)
Potential benefits	Shorten time for developing improved crops and vaccines, speed up traditional breeding, use less pesticide and chemicals, widen range of traits, increase control of research results	Decrease cost of information processing, speed information processing, increase communication, integrate research results in decision making, increase access to information

Source: Adapted from Miles (1997).

Information technology

"Information technology" is defined as advances in microelectronics based on semiconductor technology. Microelectronics have made it possible to produce, store, retrieve, communicate, manipulate, and display information in ways that are considerable cheaper and more powerful and convenient than was previously possible (Miles 1997). Agricultural research is a "taker" rather than a "maker" of information technology. In biotechnology, agricultural research is more a maker.

Information technology is speeding up and improving research because scientists have better access to information (e.g., through virtual libraries on the Internet) and they can communicate and exchange knowledge more easily and at less cost using, for example, electronic mail. Information technology also offers means to refine research planning. Examples are simulation models, which use mathematical relations to generate different scenarios for assessment, and geographic information systems (GIS), which help planners target research objectives for a particular agroecological zone. In research organization management, information technology is improving the information base for decision making. Managers can obtain, organize, and use information on resources (human, financial, physical) and activities that help them in planning, monitoring, evaluating, budgeting, and accounting.

Through network development, information technology also provides a means to bring institute staff closer or to lay contacts with practitioners from other organizations.

Where biotechnology and information technology overlap a new scientific discipline has emerged. "Bioinformatics" combines biology, mathematics, and computers. It focuses on the enormous amounts of data that are generated by researchers identifying the lengthy DNA sequences of humans, plants, animals, and microorganisms (Sobral 1999). This discipline is expected to influence the evolution of biology, because biological research is becoming inseparable from the information systems needed to support research and technological development. An example of the use of bioinformatics is the System-wide Information Network for Genetic Resources (SINGER), the genetic resources information exchange network of the CGIAR. SINGER links the genetic resources databases of the CGIAR centers and allows researchers to search for information on the identity, origin, characteristics, and distribution of the genetic resources in the collections as well as access further data on the collections (see http://www.noc1.cgiar.org).

Planning for new technologies

Use of biotechnology and information technology add specific requirements for planning because of certain characteristics of these emerging technologies, including the substantial development costs, high risk and uncertainty, the fast rate of change, the integration required with conventional programs, and limited experience with these technologies. This makes it difficult to assess the potential pros and cons of the technologies, which is an important first step in planning (Falconi 1999).

Uncertainty

One of the most important steps in planning is assessing and predicting the potential performance of technology. Uncertainty is always a factor where research is concerned. Yet for new technologies, experience is lacking and the information base is especially small, adding to the degree of uncertainty. Subjective judgments, in general, serve as the basis for assessing potential performances of new-technology projects. It is therefore crucial in planning such projects to use a planning method that reduces individual bias and incorporates technical and product knowledge.

The best known technique for eliciting subjective judgments and arriving at reliable conclusions is the Delphi method (NRC 1990). Here, experts make forecasts individually and give them to a central analyst who collates them and returns the combined forecasts to the experts, after which a new round of forecasts begins. The process continues until the participants arrive at a degree of consensus.

Another method that can be applied in situations of uncertainty is the analytical hierarchy approach (Ramanujam and Saaty 1981, Braunschweig, this volume). Like the Delphi technique, analytical hierarchy is suitable for situations in which much of the necessary data is subjective. Unique to the approach is that it recognizes bias and inconsistencies in subjective judgments. These inconsistencies can then be tested and remedied, resulting in a more consistent outcome.

Public acceptance

Public acceptance is an issue mainly related to biotechnology. Much public concern is associated with the safety of agricultural products derived from biotechnology, such as genetically modified foods and their byproducts. Addressing these concerns is now an important activity for biotechnology companies. They target consumers to convince them that their products are safe for human consumption. Scientific organizations active in biotechnology must also incorporate public awareness in planning biotechnological research. If necessary, research organizations, perhaps with government support, should prepare a campaign to educate and inform consumers about biotechnology crops and risk assessment. In addition, the benefits of biotechnology need to be explained: "Why it is being used?" And, "How it may be used in the future?" (Tabie 1999).

Organizational structure and investments

New technologies, in particular biotechnology, require considerable investments, which have at least three implications for the structure of an organization and its planning. First, in the initial phase the high fixed costs of laboratories and scientific personnel for biotechnology projects can be shared by research programs. For example, teams may shift between research on maize stem-borer and cassava mosaic virus, thus sharing the fixed costs involved in the biotechnology work. But this undifferentiated biotechnology capacity may be insufficient for developing all the expertise needed for work on a particular crop. A decision must then be made on whether a biotechnology program should pursue specialization and, if so, along what lines.

Second, given the high fixed costs of laboratories and the specialized nature of the equipment required, centralization may help ensure a higher rate of use-capacity. A major drawback of centralization, however, is that it directs biotechnology research away from locally adapted technologies. Where the size of the country allows it, applied tools such as tissue or anther culture can often be well integrated into decentralized commodity programs. New information technology further promotes decentralization, allowing researchers to be closer to the problem while applying advanced techniques.

Third is the issue of investments to be made in developing tools (what to develop in-house) versus the application of existing techniques (what to buy). The cost of tool development is quite high, even if such tools are foreseen to solve a critical problem for a high-priority commodity in a country. Since existing techniques are usually less costly, it may be more efficient to acquire these techniques, if available, for lower priority commodities. In addition, strategic alliances with public, private, and international institutions help research groups share risks and uncertainties of costly and lump-sum investments.

Some planning tools are helpful in facing the above issues. The analytical hierarchy process (see Braunschweig, this volume) may prove useful on issues of centralization and specialization, and cost-benefit analysis may help in deciding whether to develop or acquire a certain technique.

Scale of initial involvement

For any research organization investing in new technologies, a difficult issue is whether to make a major initial investment or to start with a series of small investments and pilot projects. Factors of concern here include the availability and sustainability of finances, availability of qualified human resources, the objectives for using the new technology, and the type of technology (specific or generic). For example, some developing countries embarked on biotechnology research with large investments, only to experience problems in sustaining funding levels when financial crisis hit their economy or when donor support was lost. Further, a major criterion is the specific versus generic nature of the technology (Janssen 1994). Initiating a research program to develop molecular markers for rice improvement might be readily justified, because the experience and equipment acquired can be used afterwards for other commodities. Similarly, biotechnology research projects that can use existing facilities are attractive because they require only minimal additional investment. If a project is very important, but the investments required for it cannot be used in other research programs, it may be justified to commission another, possibly foreign, organization to do the work. Regardless of whether a country aims to develop a centralized or decentralized biotechnology research capacity, in the initial stages it is wise to concentrate on only a few points. These could be linked with major commodity programs on the understanding that investments may eventually be used for other purposes.

Timing

In most planning exercises, the question entertained is whether to undertake a certain project. But in new technologies the question may be asked *when* to undertake a project – now or within the next two to five years? There are some advantages to a "late start," as costs of the necessary equipment and inputs in biotechnology and computers are falling dramatically, and scientific discovery in related areas may allow researchers to reduce uncertainty by borrowing some results from others. A danger of late starts, however, may be that findings have been patented by others.

A good reason for an "early start" is the advantage of early application of new technology, perhaps providing a country with a comparative advantage in the international market. The success of many high-value exports (e.g., horticultural and ornamental products) in the international market relies on a good information system and applications of biotechnology. In addition, gaining experience early in new technologies may open markets for the technologies; they may be sold in neighboring countries.

Another aspect of timing is that because new technologies are changing fast, their planning should be reviewed more frequently than conventional research priorities.

Lock-in

New technologies are changing so fast that acquisitions might become obsolete in less than a year, in particular in the area of information technology (software and hardware). When the costs of switching from one technology to another are significant, users face "lock-in." Understanding the costs of switching technologies is critical for recognizing and measuring the danger of lock-in in planning investments in new technologies (Shapiro and Varian 1998). In this regard, users or buyers of new technologies should plan to avoid or at least anticipate lock-in. Cost-benefit analysis following standard investment theory and analytic hierarchy process may provide important insights on the expected profitability of investing in new technologies.

Partnership models

New technologies are redefining the boundaries of public and private responsibilities in agricultural research. The private sector now conducts basic research in molecular biology and develops management information, both of which were traditionally in the public domain. Relations between the public and private sector are becoming less linear, however, for instance, through the increasing use of research contracts. New technologies require new linkage mechanisms between both sectors because of property rights, the high degree of uncertainty of outcomes, and growing private-sector involvement. Joint ventures, venture capital, and shared-risk ventures could become the most common means of partnership in the future, but this would imply a change in culture of public-sector research organizations. For example, new technologies will lead to new patterns of specialization: for example, companies may specialize in marker systems or in collecting and developing databases. Moreover, new contractual arrangements will be influenced by the changes in legal frameworks related to new technologies.

Partnerships will develop, but the outcomes are uncertain. "Scenario planning" is a planning tool that could prove useful in analyzing the uncertainty of partnerships (see Johnson and Paez, this volume). Scenario development is a disciplined method for imagining, structuring, and probing possible futures. The result of scenario development is usually a small set of alternative scenarios that highlight and contrast the different conditions that a research organization may face.

Legal framework

New technologies fall under a totally new legal framework, including intellectual property rights (IPR), for biotechnology and information technology; biosafety, for biotechnology; and privacy, mainly for information technology.

Development and enforcement of the legal framework will promote and encourage the private sector to become involved in new-technology research. Strengthening intellectual property protection for biological products and processes in developed countries has facilitated private-sector investments in biotechnology research. Technology users, including those in developing countries, increasingly have to pay for the right to use procedures or products. Rights often involve complex ownership issues, with important implications for access to products, trade, and investment. Cooperation between the public and private sectors also requires clearly communicated rules and guiding principles on IPR.

"Biosafety" is associated with the use of genetically modified organisms (GMOs). A relatively new concept in agricultural research, it tempers the adoption of a new technology by considering its potential effects on human health and the environment. Biosafety guidelines set forth policies and procedures for ensuring the safe use of biotechnology and its products. The degree to which biosafety guidelines are implemented could determine the extent to which new products are introduced in the market. Lack of resources for implementing biosafety guidelines may delay and discourage private-sector research (Traynor 1999).

The rapid increase in computing and communication power has raised concerns about privacy of personal information. As a result, confidentiality and security of personal and institutional data are now major concerns in information technology. If information technology is extensively used in the daily operations of a research organization, a privacy policy or regulations should be formulated to ensure proper handling of sensitive information.

The legal framework influences the impact of new-technology research and should be considered in the planning process. Both the analytical hierarchy process and scenario planning (see Johnson and Paez, this volume) allow implications of the legal framework on new technologies to be included in the planning exercise.

Human resources

New technologies require not only new laboratories and equipment but also new human capital: trained scientists, lawyers, managers, and information specialists. Investments in capital must be accompanied by a bolstered human resource capacity. At the planning stage, a comprehensive human resource development strategy should be drafted that includes the skills requirements for new technologies in the context of the overall corporate plan or organizational mission. The planning of human resource development should consider capacity building: capacity to formulate policies, strategies, and priorities; ability to formulate and implement a regulatory framework; skills to conduct research using new technologies; and capacity to manage new technologies. New technologies are immersed in uncertain and complex outcomes and processes, which require first-rate skills. New and more capacity is needed among decision makers, managers, and scientists to clarify policy on new technologies and the research agendas to be furthered by using them.

Diffusion models

Information technology provides new means to conduct, deliver, and diffuse research results, such as use of Internet, electronic mail, CD-ROM, and voice recognition software. Planning units should be creative in exploring and using these new means to expand the dissemination of the organization's research results. Use of information technology will spawn new branches of enterprise specialized in using this technology for marketing.

As explained earlier, some tools may be particularly appropriate for helping decision makers plan the implementation of new technologies. As most of them are explained elsewhere in this volume, this section emphasizes their contribution to dealing with some new-technology issues (table 2).

Relevance for agricultural research organizations

As the technological base of agriculture and agricultural research is changing from biological disciplines to molecular disciplines and from analogue to digital engineering, biotechnology and information technology are playing an increasing role in research. Agricultural research organizations use these new technologies to pursue their mandates, missions, and national goals (such as food security, competitiveness, and poverty alleviation).

There are many potentially beneficial applications of biotechnology in agriculture. Scientists can use biotechnology techniques to boost a plant's ability to ward off pests and disease, to improve tolerance to environmental stress, and to enhance food quality. Biotechnology can also be used to diagnose disease in animals, promote growth, and develop vaccines. Information technology can contribute to lowering the cost of technology generation by reducing the time needed to access, exchange, and disseminate information and knowledge. Both types of technologies open new research frontiers. A good example is bioinformatics, which provides a pathway to meaning in a world of complex data.

For agricultural research decision makers, three implications of using new technologies are particularly significant: the need to maintain a position in a

Table 2. Planning Tools and New Technologies

Planning Tool	New-Technology Issues Addressed
Analytical hierarchy process	uncertainty, organizational structure, lock-in, decision making
Cost-benefit analysis	investments, avoiding lock-in, timing, partnerships, diffusion
Delphi technique	uncertainty, information quality in the planning process
Scenario planning	uncertainty, partnerships

quickly moving field, the need to be an attractive international partner, and the development of cross-sectoral links.

Maintaining a position in a quickly moving field. The "position" a research institute wishes to maintain may be at the cutting edge or behind it. It is critical in planning to decide which platform is most suitable and how to build from such a position. Whereas biotechnology and information technology are now major issues on the agricultural research agenda, in 10 years' time the situation will probably be different. There will be other frontiers, with similar questions to be dealt with. Thus, there will be a need to decide what will be the organization's "growth point."

Being an attractive international partner. There are many opportunities for international collaboration in new technologies. Yet research managers must understand the challenges involved in bringing international collaboration to fruition. Some points to consider are (i) the importance of close involvement of research leaders early in the planning stage of international collaborative projects; (ii) the need to have a regulatory framework in place (in the case of biotechnology, biosafety and intellectual property regimes are integral components of international initiatives); and (iii) the need to support local capacity development to ensure optimal benefit from international collaboration (Komen 1999). Being an attractive partner leads to better access to new scientific developments.

Developing cross-sectoral links. The different techniques of biotechnology and information technology are not only useful in the agricultural sector. For example, molecular techniques can also be applied in medicine, industry, and agroindustry. GIS is applied in other sectors as well. In planning, understanding such horizontal links is critical.

Examples

The International Potato Center (CIP) studied on how to set research priorities and use the results to influence the biotechnology research agenda on the most damaging potato diseases (Collion and Gregory 1993, Ghislain 1998). In more general terms, the objective was to set priorities among projects within a commodity, refine resource allocation guidelines, build consensus on the research agenda and funding allocations, and put in place a system for ongoing priority setting. Emphasis was squarely on translating priorities into guidelines for resource allocation. Scoring combined with a simple cost-benefit model was chosen for the priority setting. The exercise took about six months and was finished in 1992. CIP management and scientists participated in priority setting. The results of the exercise were translated into project scores in which research on late blight emerged as CIP's top research priority.

Shapiro and Varian (1998) provide economic principles for planning use of information technology and distill them into practical strategies, cases, and guidelines.

References

Cohen, J. I. 1994. Biotechnology Priorities, Planning, and Policies: A Framework for Decision Making. ISNAR Research Report No. 6. The Hague: International Service for National Agricultural Research.

Collion, M. -H. and P. Gregory. 1993. Priority Setting at CIP: An Indicative Framework for Resource Allocation. Lima: International Potato Center.

Falconi, C. 1999. Methods for priority setting in agricultural biotechnology research. In *Managing Agricultural Biotechnology: Addressing Research Program Needs and Policy Implications* edited by Joel I. Cohen. Wallingford, UK: CAB International.

Ghislain, M. 1998. El enfoque del CIP en la implementación de una estrategia nacional para el desarrollo de la biotecnología. In *Transformación de las Prioridades en Programas Viables: Actas del Seminario de Política Biotecnólogica Agrícola para América Latina* edited by J. Komen, C. Falconi and M. Hernandez. The Hague: Intermediary Biotechnology Service.

Janssen, W. 1994. Biotechnology priority setting in the context of national objectives: state of the art. In *Turning Priorities into Feasible Programs: Proceedings of a Regional Seminar on Planning, Priorities and Policies for Agricultural Biotechnology in Southeast Asia* edited by J. Komen, J. I. Cohen, and S. K. Lee. The Hague/Singapore: Intermediary Biotechnology Service/Nanyang Technological University.

Komen, J. 1999. International collaboration in agricultural biotechnology. In *Managing Agricultural Biotechnology: Addressing Research Program Needs and Policy Implications* edited by Joel I. Cohen. Wallingford, UK: CAB International.

Miles, I. 1997. Contemporary technological revolutions: characteristics and dynamics. In *New Generic Technologies in Developing Countries* edited by M. R. Bhagavan. Stockholm and London: Swedish International Development Cooperation Agency and Macmillan.

NRC (National Research Council). 1990. Plant Biotechnology Research for Developing Countries. Washington, D.C.: National Academy Press.

Ramanujam, V. and T. Saaty. 1981. Technological choice in the less developed countries: an analytic hierarchy approach. *Technological forecasting and social change,* 19: 81–98.

Shapiro, C. and H. Varian. 1998. Information Rules: A Strategic Guide to the Network Economy. Boston: Harvard Business School Publishing.

Sobral, B. 1999. Bioinformatics and the future role of computing in biology. http://www.ncgr.org.

Tabei, Y. 1999. Addressing public acceptance for biotechnology: experiences from Japan. In *Managing Agricultural Biotechnology: Addressing Research Program Needs and Policy Implications* edited by Joel I. Cohen. Wallingford, UK: CAB International.

Traynor, P., 1999. Biosafety management: key to the environmentally responsible use of biotechnology. In *Managing Agricultural Biotechnology: Addressing Research Program Needs and Policy Implications* edited by Joel I. Cohen. Wallingford, UK: CAB International.

Chapter 5
Planning, Performance, and Accountability

Warren Peterson

Organizational performance and accountability are increasingly important for public-sector agricultural research organizations. Pressure to improve performance and accountability comes from a variety of stakeholders, including donors and clients. Calls for accountability have led to a search by agricultural research system managers, as well as investors and consultants, for practical, sound means of assessing, demonstrating, and improving research organizations' performance and results. This chapter looks at the need for improved performance and accountability. It presents options for how organizations can improve these aspects of their operations, referring particularly to questions of how performance and accountability can be addressed in a planning context.

What is organizational performance and accountability?

"Performance" refers to an organization's ability to plan and use resources to produce outputs that are relevant and useful for its target users or clients. A performance-oriented agricultural research organization is necessarily focused on producers/users and on research output productivity. This definition of performance highlights two important dimensions: one related to productivity and outputs and the other related to the relevance of these outputs for the organization's stakeholders. This latter aspect is closely linked to the idea of accountability.

While the nature of its outputs depends on the organization's mandate, and specific outputs can be defined and identified in different ways, any research organization needs to produce products that are relevant to users. Furthermore, to justify the investments required to maintain operations, the organization needs to produce those outputs in an effective and efficient manner. Here "effectiveness" means an organization's ability to produce outputs that correspond to its goals and to users' needs; "efficiency" is the achievement of planned outputs using a minimum of inputs. Performance is related to planning through the need to identify, measure, and evaluate outputs in terms of planned objectives and the resources used to achieve them.

Accountability and performance are linked because public-sector agricultural research organizations must account to their investors (government and donors) for their use of funds and to their users (producers, extension agencies, actors in the agricultural knowledge system, and units of government) for the appropriateness of the technology and information they have generated. Accountable management implies that individuals and organizations are responsible for specified levels of performance (Premchand 1993).

Public-sector organizations in developed and developing countries are under pressure to improve their performance in terms of enhanced productivity and accountability, and agricultural research organizations are no exception. There are four main reasons why issues of performance and accountability have gained prominence. First, governments have been affected by structural adjustment and a shrinking public sector. Second, in many countries ideas from "the new public management" have led to calls for more productive and entrepreneurial government that focuses on results rather than red tape. The new public management has led to reassessment of boundaries between the public and private sectors. Third, in many countries there is disappointment with the performance of agricultural research organizations, as the "easy" productivity gains from investments in research, technology transfer, infrastructure, and rural support services have largely been realized. Finally, developing countries tend to move towards more open and democratic forms of government. In this context there is an increased demand for the public sector to be accountable and relevant to its clients and stakeholders (Lusthaus et al. 1995).

Most investors perceive as weak the performance of public-sector agricultural research organizations in delivering outputs, particularly in developing countries (Byerlee and Alex 1998). Worsening conditions for groups that are socially and economically vulnerable coinciding with declines in per capita income, widening trade imbalances, and growing external debt, are apparent in many countries and regions. Also, exaggerated expectations of what a research program might accomplish have led investors to lose confidence in the ability of public-sector agricultural research to contribute to improving the situation. The result has been a general pattern of decreasing investments that has affected the stability and effectiveness of national research organizations.

The ability to address these performance and accountability issues will in part determine future investments in public-sector agricultural research, and ultimately the implementation and realization of public-sector agricultural development policies and objectives. Yet in many agricultural research organizations, managers are uncertain about what can be done to improve performance and accountability or how to proceed in doing so.

A central problem is that current management systems and attitudes are seldom centered on performance. Nor do planning, monitoring, evaluation, and reporting methods reflect a performance orientation. One basic requirement to improve performance is therefore to develop or alter management systems and practices so that they focus on research outputs, results, and their relevance for users. Internally driven evaluations offer the chance to combine the evaluation goals of accountability with performance improvement. Until agricultural research organizations adapt their management and evaluation systems so that performance and accountability are the guiding elements, improvements in performance are unlikely.

Current evaluation practice in agricultural research is dominated by donor project evaluation procedures. These, however, address individual projects rather than the performance of the organization as a whole. In addition, evaluations are often performed using externally driven methods and personnel. As a result, evaluation in many research organizations is aimed at external accountability and improvements in managing donor projects, rather than the management of the organization.

Evaluation approaches commonly used in other public sectors (e.g., education and law enforcement) and by commercial enterprises are almost unknown in agriculture. These approaches are aimed at performance evaluation and improvement and offer various means of bettering the perspectives and practices used in agricultural research (see CCAF 1993, Lusthaus et al. 1995). A combination of quantitative and qualitative methods are hereby applied for the analysis of outputs and their implications for users, linked with some means of evaluating key management tasks.

Performance-oriented management systems

Establishment of performance-oriented systems that serve the organization is an important step for agricultural research managers. Building a system that focuses on internal management and performance sends a powerful message to investors that their concerns are taken seriously and that performance improvement is a key management objective. A performance-oriented management system has to be tailored to the specific needs of the organization. For example, the US National Science Foundation (NSF), in response to the Government Performance and Results Act (GPRA), designed a performance assessment system that consists of three main elements: strategic planning to specify goals and objectives, databases of outputs, and periodic external evaluations (NSF 1995).

A number of components and activities contribute to a performance-oriented management system. First are procedures for analyzing, assessing, and measuring the organization's outputs. Most public agricultural research organizations produce far more technology, service, and information outputs than are recognized by either their national staff or their clients and investors. Managers need to identify outputs, determining productivity by measuring inputs compared to outputs, and successfully communicate them to investors and stakeholders.

Second are methods for determining the relevance and benefits of research outputs for the organization's clients (primarily farmers and other units of government). Survey methods are most appropriate, and the results can be used in making the organization more responsive to clients' needs.

Third is periodic strategic planning in which strategies are defined to operationalize government policies and development objectives and to set the basic directions and objectives of research. The organization's performance in terms of these policies and objectives can be examined for the fit between the strategies and planned outputs, as well as for impact on governmental economic and development goals.

Fourth, effective program planning is needed to implement the strategic plan and identify a rational portfolio of research projects to be undertaken during a specific period. Program planning procedures should include planning at the component project level. Here, it is crucial that project proposals and logical frameworks define objectives and outputs – and inputs (resources) needed to produce the outputs – as well as identify verifiable indicators of their achievement. These allow output measurement and project monitoring.

Fifth, internal and external reporting and evaluation cycles should be developed. A principal means of improving accountability is good communication and reporting to external stakeholders. These rest on the establishment of reporting cycles wherein information on progress is provided to different management levels within the organization and to external stakeholders. Periodic internal program reviews or external evaluation may complement such reporting.

Finally, performance-oriented management needs to be enhanced by identifying and improving weaknesses in management. Since good management is directly related to the efficient production of effective research outputs, research organizations need a method for assessing management. Such assessments can provide a basis for correcting management problems and thereby improving performance.

A performance-oriented management system can be built in different ways. One approach is presented in Peterson (1998). Whatever the approach used, elements related to planning and reporting are particularly important for improving performance and accountability.

Improving performance and accountability

Managers must approach the issue of organizational performance from two perspectives. The first is an internal "performance" perspective that examines the organization from the point of view of output productivity and improvement of management. The second is an external "accountability" perspective that examines the organization from the viewpoint of users, investors, and other stakeholders.

The performance perspective

Better research organization performance can be achieved by planning and assessing outputs (in relation to plans, available resources, and user needs) and by establishing adequate systems and means of evaluating, measuring, and reporting results. Performance issues can be addressed at both the strategic planning level and the level of program planning. Strategic planning provides a framework in which to plan programs and projects that conform to national policy goals and government-set research objectives. Program planning defines the portfolio of projects that an organization will undertake within a specified time frame.

A research organization's performance may be assessed at two levels. The first is that of the outputs and outcomes it produces. The second is assessment of critical management factors that drive the organization's performance. With regard to outputs and outcomes, there are several means of integrating performance concerns into research planning:

- identifying, measuring, and assessing an organization's outputs in relation to its mandate, strategy, and objectives
- designing project plans and logical frameworks that specify objectives and outputs responding to client needs
- developing indicators of success and achievement for inclusion in logical frameworks and for use in project monitoring and reporting
- measuring output productivity
- designing procedures for tracking output use by farmers

The second approach to measuring and improving performance – reviewing the key management domains that determine an organization's capacity to produce outputs and outcomes – can be accomplished byseveral means:

- assessing management performance in critical areas
- identifying management factors that need improvement
- implementing measures to redress management weaknesses

The accountability perspective

The perceptions of investors, users, and other actors need to be considered when addressing accountability. Such "views from without" gauge how successful the organization has been in conveying information about its performance to those outside. Accountability also entails transparency in making decisions, setting objectives, formulating plans, and evaluating results.

Both strategic and program planning, if carried out with suitable stakeholder participation and transparency, are effective instruments for communicating accountability. These processes offer stakeholders opportunities to contribute to planning, and the documents produced provide a transparent and detailed source of information about research objectives and the resources needed to achieve them.

In the short run, accountability may be enhanced through information, communication, and participation mechanisms that satisfy the partners in research. Some means to achieve this can be named:

- building a transparent management system that is focused on performance
- ensuring the participation of stakeholders and users in the various management processes used in planning, decision making, and evaluation
- open, objective reporting and communication about the use of funds and the results obtained

In the longer run, improved accountability can be attained through careful planning and management of linkages with a variety of stakeholders. Linkages become increasingly important when public-sector research organizations start focusing on client needs, when they become involved in public-private partnerships, and when they diversify their sources of income. Maintaining formal linkages is costly and time consuming; careful planning and design is needed to reap benefits. Perhaps the four most important aspects of planning linkages are identifying linkage needs and partners, specifying linkage functions, defining linkage mechanisms, and determining resource requirements (Peterson et al. 1999).

The two main *linkage needs* are to establish technology transfer and information flows so that research activities and outputs can be made available to others, results and impact can be communicated back to research, and coordination and planning among actors can be done effectively. Accountability relates specifically to information flows. Appropriate *linkage partners* need to be identified to cover the different needs. These include farmers and farmers' organizations, extension services, other public-sector research organizations, international and regional organizations, nongovernmental organizations, the private sector, donor and development agencies, and government regulatory and policymaking bodies.

There are two basic *linkage functions*: to plan and coordinate activities between the research institute and other organizations involved in technology generation and transfer, and to establish two-way flows of technology and information among the research organization, end users, and other actors. For each of the actors identified, specific functions and purposes can be listed. Linkages with farmers, and farmers' organizations, for example, serve to provide information on farmers' technology needs, to contribute to research planning and review, to channel technology and information to farmers, to provide feedback on the usefulness of research outputs, and to stimulate farmer participation and farmer-to-farmer dissemination.

The linkage functions can be realized through a variety of *linkage mechanisms*. Planning these mechanisms should reflect the functions identified and be realistic in terms of resource requirements. Linkages are expensive and must be budgeted on an annual and medium-term basis. They are a major budget expenditure, and if they are not explicitly included in costing, they will not be implemented. Priority rankings of linkage functions and mechanisms may be needed due to their high costs.

For each mechanism defined earlier, *linkage resource costs* must be estimated. These include travel (e.g., fuel, per diem), materials, equipment, and staff time. In addition, responsibilities for managing each mechanism should be delegated.

Relevance for agricultural research

Public-sector agricultural research organizations need to respond to demands for improved performance and demonstrate both their relevance and effective and efficient use of funds. In many research organizations, these needs are not adequately met. Ignoring these demands, however, can lead to significant consequences, including reduced funding and loss of confidence in the research organization by stakeholders and users.

Organizations can reap significant benefits by establishing a performance-oriented management system and a performance-assessment system that improves both organizational performance and its ability to communicate its accomplishments externally. Planning and managing performance and accountability provides a basis for

- focusing on research productivity in terms of outputs and results
- improving performance in terms of relevance to users
- linking planning to objectives and outputs
- identifying and correcting management problems
- improving communications and transparency
- enhancing institutional sustainability for public-sector research organizations

To gain these benefits, research organizations must establish internal management processes that focus on research results, plan research objectives and outputs, define indicators of success, and then monitor, adjust, and articulate results.

Examples

In the Palestinian Authority, partners and stakeholders in agricultural research participated in strategic planning for government research. The transparent process resulted in broad stakeholder agreement on issues, priorities, and strategic objectives for agricultural research. This was followed by program planning, in which a portfolio of research projects to be undertaken over a five-year period was identified. The projects were designed to achieve the priority objectives that were identified in the strategic planning exercise.

Each component project was planned using a logical framework approach that identified specific outputs and established verifiable indicators for their achievement. The time frames, activities, and resources for each output were also indicated to aid in project monitoring. Procedures and responsibilities for monitoring and adjusting projects in terms of outputs were then established and internal and external reporting responsibilities defined.

Agricultural research now reports regularly to government, donors, and stakeholders, communicating progress made towards project outputs in relation to the predefined plans and indicators of achievement. Transparency and participation in planning, coupled with objective progress reports, have improved the ability of the research organization to demonstrate its performance.

Cyprus

Accountability and performance in terms of research activities should be complemented by performance assessments of research organizations as a whole. Such an overview serves donors as well as national interests. In Cyprus, an organizational performance assessment system was established with this objective in mind. The system allows the agricultural research institute to conduct periodic self-assessments of performance and identify areas of management that can be improved.

The first step is an assessment of research outputs. Stakeholders defined six categories of output: recommendations on improved breeds and crop varieties, publications and reports, dissemination events, crop and livestock management practices, training events, and public services. Outputs in each are quantified for the years covered by the assessment. Productivity measures (output divided by input of researcher time) are applied to establish an index performance ratio for each category.

The second step is to assess key areas of management, because the quantity and quality of research outputs is directly affected by the effectiveness of management. A working group examines 10 key areas of organization management: assessing context, planning strategy, selecting objectives, managing projects, maintaining research quality, ensuring staff quality, coordinating internal functions, transferring technology, protecting organizational assets, and ensuring information flow.

Within each key area, management processes and procedures are scored to determine the extent to which they are used and improved. The scores for each element, process, or procedure in a key area are arrived at by working group consensus and discussion. Factors of management that could be strengthened are also identified in each area. The cumulative score for each key management area is then divided by the total possible score to yield a performance ratio. The ratio is used as a benchmark for future assessments.

External constraints on management are further identified in each key area and a constraint ratio calculated. Such constraints can greatly affect research outputs. This step allows managers to differentiate elements for improvement that can be addressed by the organization's management from those imposed by outside factors.

The research output and key management results are then compiled and discussed. Conclusions and recommendations for action are finally agreed upon and an assessment report prepared for management.

References

Byerlee, D. and G. E. Alex. 1998. Strengthening National Agricultural Research Systems – Policy Issues and Good Practice. Washington, D. C.: World Bank.

CCAF. 1993. Reporting and Auditing Effectiveness. Ottawa: Canadian Comprehensive Auditing Foundation.

Lusthaus, C., G. Anderson, and E. Murphy. 1995. Institutional Assessment: A Framework for Strengthening Organizational Capacity for IDRC's Research Partners. Ottawa: International Development Research Centre.

National Science Foundation. 1995. Performance assessment at the National Science Foundation – Proposals for NSF's response to the Government Performance and Results Act. Washington, D.C.: National Science Foundation.

Peterson, W. 1998. Assessing Organizational Performance Indicators. The Hague: International Service for National Agricultural Research.

Peterson, W., M. Wilks, and A.Wuyts. 1999. Linkages between Research, Technology Transfer, and Farmers' Organizations. Research Report No. 15. The Hague: International Service for National Agricultural Research.

Premchand, A. 1993. Public Expenditure Management. Washington, D.C.: International Monetary Fund.

Part II
The Content of Agricultural Research Planning

Helen Hambly Odame

Research planning has assumed many different forms (Hambly and Setshwhaelo 1997). Plans differ in their object (national, institute, program, or project level). They may address different types of resources (human, financial, or infrastructure) or encompass a specific time frame (long, medium, or annual terms). Plans also may be characterized as either more strategic or operational in nature, involving a comprehensive or targeted focus for resource allocation. The six chapters in this part of the sourcebook present the range of plans most often associated with agricultural research. Without recommending a "fixed menu" of planning types, each of the chapters identifies the key characteristics of a plan type, the major steps involved in its formulation, and its relevance to agricultural research organizations.

The types of plans and examples presented here are those found in public-sector agricultural research. That said, it must be recognized that the private sector also engages in activities such as strategic planning and foresight studies. From this review of agricultural research planning, a general typology of plans can be distinguished, although planning types often merge or overlap both in time and in function.

Typology of research plans

Agricultural research plans may have very different purposes, focusing on different objects, and spanning different time frames (table 1).

No agricultural research organization is capable of undertaking all these different types of planning at the same time. Given the variety of different planning types and the costs involved, agricultural research organizations need to select carefully when and how to engage in different planning activities. In particular, they need to

- make optimal use of limited planning capacity
- recognize that anticipating the future becomes more difficult in a dynamic external environment
- identify the right participants for the different types of planning
- match information requirements to planning type
- avoid excessive emphasis on planning at the cost of implementation.

Table 1. Types of Agricultural Research Plans

Type of Planning	Purpose	Object	Time Frame
Foresight studies	Explore future dynamics of policy, science, and technology	Sector	5–15 years
Agricultural research policy development	Develop a framework to guide agricultural knowledge and technology generation	Sector	4–10 years
Master planning	Define long-term investments and activities	Institute	5–10 years
Strategic planning	Identify need for and direction of change	Institute	4–8 years
Program planning	Focus research on priority constraints within program domain	Institute or program	3–5 years
Project planning	Develop an efficient and sufficient set of activities to overcome a constraint	Institute or program	1–3 years
Experiment planning	Develop the best option to obtain insight in a scientific question	Institute or program	1 year or less
Financial planning	Match financial availability and needs	Institute	Variable
Training planning	Develop human resources	Institute	Variable

Any simple classification of research plans by subject or time frame, such as that in table 1, is complicated by the recognition that planning is also characterized by the degree to which it is strategic or operational in nature. The extent to which plans capture the detail needed to determine broad investment schedules or targeted resource planning is also a distinguishing factor. Furthermore, there is a distinction between the types of plans that focus specifically on planning research and others that consider in-depth the need for institutional development.

Strategic versus operational planning

The role and importance of planning often hinges on the capacity of research plans to respond to new and emerging challenges. Identifying these new directions is the focus of planning that is strategic in nature and typically involves a comprehensive analysis of internal and external strengths, weaknesses, opportunities, and threats.

In planning literature there has been some debate on the role of strategic planning. While some experts in this field view the role of strategic planning positively (e.g., Bryson 1995, Galagan 1997), others argue that strategic planning exercises should be redesigned in favor of a more dynamic approach to reacting to changes in the external environment. The important consideration in this debate is whether strategic thinking can improve research priorities and resource availability, eventually leading to positive changes in the sector.

Without doubt, agricultural research organizations must ensure that they maintain a clear strategic outlook. An understanding of agricultural research policy and, possibly, conducting foresight studies, can help research organizations identify their strategic direction. Any planning effort must analyze new funding mechanisms and find innovative answers to other challenges such as changing producer and consumer demands and new technologies.

Nevertheless, to respond to new challenges, operational types of planning cannot be neglected. These include program, project, and experiment planning, which in some research institutes may be set within the framework of a medium-term plan. In this respect strategic and operational planning can be linked so that one is articulated in relation to the other. It would be hazardous for instance, to embark on a medium-term planning process without at least a general sense of the strategic issues and priorities facing the organization. Yet ISNAR's experience in national agricultural research planning suggests that operational planning has sometimes occurred without the sense of direction that strategic planning conveys. The reverse may also be true: due to over-emphasis on the strategic side, the operational aspects of planning are delayed or postponed. Segregating these two types of planning undermines planners' ability to relate activities within the organization to broader developments in the agricultural research system to which the organization belongs.

Investment planning versus resource planning

Agricultural research and development in many countries has been acutely affected by resource fluctuations. It is in this variable climate that agricultural research organizations find themselves faced with long-term investment decisions in new areas, such as biotechnology and sustainable development.

One mechanism for guiding longer term investment decisions is master planning. While master plans are intended to be strategic in nature, they also include some degree of operational planning in order to link resource allocation to agricultural research priorities. A distinguishing aspect of master plans is that they are mostly used for making or channeling major financial investments – not for redefining an organization's position vis-à-vis its environment.

Other forms of resource planning can take place alongside investment or master planning. Special plans for finance, human resources, and infrastructure are common in agricultural research. Funding strategies and financial plans are becoming increasingly important to agricultural sectors in developing countries due to reductions in overseas development assistance as well as new partnerships emerging between the public and private sectors. In addition, the rapid internationalization of research staff has meant that more attention must be paid to planning human resources, and training is fundamental to ensure that past and present efforts to build capacity in agricultural research remain uncompromised.

Institutional development versus research planning

Certain types of research plans, such as master plans, address the need for comprehensive institutional development in the agricultural sector. Master planning is comprehensive, defining the structure, policy, strategy, program, and resource requirements of a research organization. In some cases, a master plan may lead to the creation of a new institution. Because it is so ambitious, master plans' implementation may be constrained by lack of funds or by rapid changes in the external environment. In contrast, planning can be less ambitious, accepting the existing institutional framework as its starting point. In such cases, emphasis is placed on planning research activities, and not on the organization and structure that supports the research.

Sequencing various research plans

The sequencing of agricultural research planning has not been addressed adequately in the literature or in research management. Yet sequencing of planning has often been a point of contention, if not confusion, among managers. Planning models that suggest a linear progression from policy to strategy to medium-term and annual plans is incompatible with the reality of agricultural research. Diversity in demands among national agricultural research organizations and their stakeholders (i.e., clients, investors, and government) leads to a diversity of approaches to planning. And while developing a plan on a one-off basis is of limited use, organizations are wary of engaging in constant planning and replanning of their work. Research managers may feel there is a surfeit of planning efforts in their national research system.

More recent models of planning have moved away from the linear approach. The new emphasis is on responding to changes in the environment and positioning research strategically amidst new alliances and niches (Bryson 1995). This shift to a more dynamic model still values planning, but does not insist upon a rigid chain planning at the strategic, program, and project levels. Research priorities are still identified, but effort is placed on the organization taking action towards attaining these priorities and establishing its strategic niche.

The chapters

Research plans need to reflect the country's agricultural research policy. Therefore, the first chapter in this part highlights the core elements of *research policy development*. Research policy provides a framework within which decisions and investments are made and activities implemented in agricultural research. Policy informs the planning process and its deliverable: the research plan and its implementation.

Exploring long-term developments that go beyond the foreseeable "planning horizon" is the domain of *foresight studies*. Yet foresight methods are relatively new in agricultural research. They are used to develop a joint perspective on the future and involve multiple new partners and stakeholders.

Strategic planning provides a conceptual framework for decision making. It directs attention to intra-organizational understanding and cooperation and provides a means of coping with and surviving turbulence in the environment. The latter is typically accomplished by rethinking the organization's mission, identifying strategic issues, moving significant resources from lower to higher priority programs, improving organizational morale and image, and identifying a comprehensive and proactive approach to research management.

Master plans tend to respond to specific policy requirements of investors and governments and emphasize the financial and human resource implications of the organization's development. They are more useful for incipient organizations than for mature ones. The comprehensive process and product of master planning have tended to be referred to as a "blueprint model" of long-term planning, often facilitated by an external team of specialists.

Program planning refers to the process by which research programs are broadly defined for the medium to long term. They implicate human, technical, and financial resources within a particular domain or subsector. Generally, each research program is made up of a number of projects. *Project planning* concentrates on integrated and finite activities designed to resolve one or more problems detected in a subsector. Its problem orientation requires specific objectives to be identified, a proposal prepared and reviewed, and finance found for implementation. Monitoring and evaluation are integral parts of the project planning process.

Experiment planning is the lowest level of planning addressed in this sourcebook. The experiment proposal details prospective research. Consideration of influential factors such as natural cycles, financial regulations, deadlines, and higher level program or project implementation is essential.

In addition to the various types of research plans, there are specific resource-based plans focus on key elements of agricultural research including human resources, infrastructure, and finance. *Financial planning* aims to reconcile the level of research activity with the likely availability of funds from different sources.

Planning human resources and, specifically, training has become critical in recent years as greater efforts are made to ensure capacity is built in agricultural research. *Training planning* identifies the knowledge, attitudes, and skills required to fulfill the mission of the organization and to ensure its responsiveness to emerging research programs and priorities.

Diverse processes support the various types of plans devised in agricultural research. The major processes of developing and implementing agricultural research plans are discussed in detail in Part III of this sourcebook. Like the external policy context in which plans emerge, the planning process shapes the resulting plan, its implementation, and inevitably its relevance and effectiveness.

References

Bryson, J. M. 1995. Strategic Planning for Public and Nonprofit Organizations: A Guide to Strengthening and Sustaining Organizational Achievement. San Francisco: Jossey-Bass.

Galagan P. A. 1997. Strategic planning is back. *Training & Development*, April: 32–37.

Hambly, H. and L. Setshwaelo. 1997. Agricultural Research Plans in Sub-Saharan Africa: A Status Report. Research Report No. 11. The Hague: International Service for National Agricultural Research.

Mintzberg, H. 1994. The Rise and Fall of Strategic Planning. New York: Prentice Hall.

Chapter 6
Agricultural Research Policy Development

Steven Were Omamo, Michael Boyd, and Willem Janssen

An agricultural research policy is a framework for investments and activities in generating and disseminating agricultural knowledge and technology. It is developed in a process that links innovation in agriculture to prospects for growth and development within the agricultural sector and in the broader national economy. Attaining agricultural research policy objectives improves the likelihood that agricultural and national policy objectives will be met. Regardless of whether an agricultural research policy framework is made explicit, it will exist in the form of the extant structures and processes. The more coherent and transparent the framework, the clearer the signals will be that it sends to agricultural research organizations as they develop their strategic plans and set research priorities.

What is an agricultural research policy?

An agricultural research policy is the framework in which decisions and investments are made and activities implemented to generate and disseminate agricultural knowledge and technology. It defines what categories of knowledge and technologies should be generated and diffused in order to achieve sustainable agricultural development. Further, it tells how and by whom outputs are to be produced (Idachaba 1996). As with all planning, policy development is a process comprising design and implementation stages. But because the returns to developing a national research policy accrue throughout a country's agricultural research system, research policy development is fundamentally different from strategic planning, master planning, or any of the other kinds of planning described in this book. No single organization – private or public – has the incentive or ability to take on sole responsibility for the task. Research policy development thus must be initiated and facilitated by government. Developers of agricultural research policy do not focus on the conditions facing a particular organization or group of organizations. Rather, based on assessments of the potential contributions of agricultural research to national development, they take a broad view that spans several sectors in an economy.

Developing an agricultural research policy

A research policy is a public good. A fundamental question in its development is thus how deeply involved will the public sector be in agricultural technology development and dissemination. This question brings to light trade-offs that must be made in the policy development process: more public-sector involvement might mean a less active private sector, with important consequences for returns to investments in various segments of a research system. But there also might be public/private complementarities to explore and exploit: returns to private-sector investment in agricultural technology development may hinge on public-sector presence; and public initiatives may rely on private investment for their sucess.

These considerations should guide the choice of stakeholders who provide input, either informally or formally, in task forces to develop the research policy. Contributors in the overall process should include representatives of key government entities (e.g., ministries of agriculture, rural development, finance, planning, and environment; publicly supported agricultural research organizations; universities; and extension services) and the private sector (e.g., farm input and machinery firms, commodity processors, wholesalers and retailers, financial institutions, and farmers). Among them, they should possess a thorough knowledge of the agricultural sector, an understanding of national development objectives and policies, and an awareness of the potentials and limitations of research as an instrument of economic and social policy. Stakeholders may well vary in the different stages that comprise the agricultural research policy development process.

Figure 1 presents a useful road map for agricultural research policy development. It synthesizes issues raised by Elliott (1996) and Idachaba (1996) (see also ISNAR 1990, 1997b). Five components, called "stages" here, are elaborated below.

Stage 1. Specify the country's philosophy of development and national development objectives. A development philosophy springs from a society's values, beliefs, and aspirations. Typically, these are concretized in the economic and social policies pursued by a government. Development objectives reflect this philosophy and often are translated into quantitative macroeconomic and other sectoral targets. The processes by which a given country develops and expresses its philosophy and overarching objectives vary, but as the top-level position in figure 1 suggests, these will generally be the domain of the top level in the political hierarchy, for example, the presidency, the cabinet, and the parliament.

Stage 2. Assess the factors impeding or promoting attainment of development objectives. These factors may relate to the structure of production in a country, demographic features that influence patterns of consumption, and institutional conditions that influence terms of trade and exchange. This stage and the next

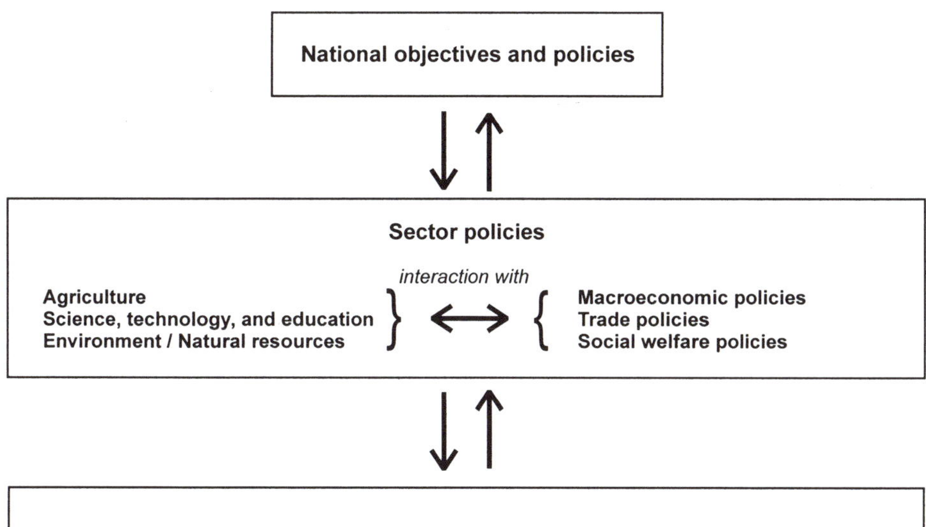

Agricultural research policy process

Area	Issues	Tools
Set research agenda	Priority setting	Strategic planning and priority setting
Funding strategy	Public-private mix	Priority setting and funding mechanisms
Knowledge dissemination	Extension and education	Participatory needs assessment, OFR, diffusion mechanisms
Links among NARS elements	Structure and coordination governance	Information systems and exchange, joint planning, incentives for interaction
International linkages	International networks	Information systems, regional collaborative mechanisms, bilateral programs
Legal framework	Intellectual property rights and biosafety	Legislation, research protocols

Figure 1. Links between the agricultural research policy process and the national policy framework

Source: Elliott 1996, ISNAR 1990, 1997b.

entail the translation of broad development objectives into concrete and implementable policies for the various sectors of the economy. In its purely public-sector respect, it is the work of line ministries, which make specific budget decisions and allocations. Other stakeholders, including the private sector, incorporate the broad sectoral directions of public policy as well as their own expectations of other institutional and environmental factors in making their choices.

Stage 3. Identify the role of agriculture in national development and specify the agricultural policy objectives that spring from this role. The share of agriculture in gross domestic product, the spatial configuration of agricultural production, and how these have changed over time are indicators of the potential role of agriculture in national development. Agricultural policy reflects this role by translating macroeconomic objectives into desired patterns of agricultural production and trade and by specifying how the attainment of agricultural policy objectives is contingent upon foreseen macroeconomic conditions. The greater the role of agriculture in an economy, the wider should be the link between sectoral and national objectives.

Stage 4. Specify the potential contribution of agricultural research to achieving agricultural policy objectives. Technological change is just one agricultural policy instrument among many. (Some of these others are market support, credit, infrastructure, education, income support, and legislation.) It is particularly suited for making long-term changes in productivity and comparative advantage. The technical potentials of alternative research thrusts define the boundaries of agricultural research impact on growth in the agricultural sector and in the wider economy. This stage is linked to all of the six elements of the agricultural research process in figure 1. It is described in more detail under stage 5. This stage is especially important in setting the research agenda and funding strategy. Because these embody issues of priority setting, the activity mix, and emphasis of public- and private-sector activity, the nature of stakeholder participation in defining policy objectives in these areas is critical. Participation must be as broad-based and effective as possible, to ensure that the complementarities mentioned earlier are fully taken on board.

Stage 5. Identify agricultural research policy objectives by developing and setting objectives in each of the following component areas:

- an agenda for agricultural research
- a strategy for funding agricultural research
- a strategy for international acquisition and exchange of agricultural innovations and information
- structures and processes for interaction among domestic agricultural research institutions
- a strategy for agricultural technology dissemination within the country
- a legal framework that supports agricultural technology development and dissemination

Simply listing these component areas underscores the multi-faceted nature of the process. The critical factor is that well designed and fully participatory processes for each area are put in place. Involvement of all relevant stakeholders increases the likelihood that the outcome will take account of connections among stakeholders and areas and engender commitment to implementation.

The relative importance of each stage in the framework differs among countries, depending on the constraints facing agricultural technology development and dissemination. But in all cases, each stage should reinforce the others. Together, they should complement the broader policy frameworks into which the agricultural research policy fits (e.g., those pertaining to the national economy, the agricultural sector, the environment, and the state of science and technology in the country).

Implementation presents its own issues, because the outcomes of policy design and implementation are jointly determined. A policy framework that is difficult to implement is likely to have been poorly designed, and vice versa. Good prospects for implementation are likely to go hand in hand with broad representation and participation of stakeholders and deeper level of consensus among them during policy design. Still, even the best designed policies will fail to achieve their objectives if improperly implemented.

Agricultural research stakeholders within a country take as given the broad national agenda and its articulation into sectoral policies. An agricultural research policy framework specifies what these stakeholders believe is the potential contribution of agricultural research to achieving these goals. But, as noted, agricultural policy development is a process. As changes occur in the environment within which policy is being implemented, the policy too must be adjusted. Flexibility is therefore an asset. But since the research policy sets the principles and rules that constrain, direct, and govern behavior in a research system, so too are stability and predictability desirable. A policy framework in constant flux undermines the rules and guidelines embedded within it. There is thus a trade-off between flexibility – the ability to adjust to changing circumstances – and stability and predictability – the capacity to retain form and substance despite alterations in the external environment. This trade-off can be resolved, through a transparent and inclusive process of policy design. Such a process ensures stakeholder commitment to the outcome. It improves stability of the policy framework and expands scope for flexibility in implementation.

Relevance for agricultural research

Regardless of whether an agricultural research policy is made explicit, it does exist, embodied in the structures and processes that define opportunities and constraints facing a country's agricultural research system. Coherent and transparent frameworks send clear signals to research organizations as they develop their strategic plans and set research priorities. Coherence and transparency, thus, lead to more effective and efficient resource allocations in terms of their contributions to national development objectives. Conversely, an incoherent, opaque policy framework sends confused signals to research actors, which diminishes agricultural research's potential contributions to national growth and development. Most important, these positive and negative consequences can be mutually reinforcing. Clear, relevant research policy frameworks are likely to lead to more successful strategic plans and priorities being developed under their ambit; and these, in turn, lead to better investments made and priorities set in the context of these plans. This virtual cycle is closed by the policy framework itself appearing insightful and reasonable in hindsight, due to its tangible outcomes. The converse is also true. An incoherent and inappropriate policy framework increases instability within a national research system, erodes the sustainability of its component organizations, eventually confirming the incoherence and inappropriateness of the policy framework.

Examples

Countries' efforts to develop agricultural research policies have differed significantly. Benin, for example, made its agricultural research policy framework explicit through a process very similar to that described earlier in this chapter. The process was initiated and facilitated by Benin's main public-sector agricultural research organization, the Institut National des Recherches Agricoles du Bénin (INRAB). The quality of the policy design process – and thus prospects for successful policy implementation – was optimized by a sequence of actions:

- involving a broad-based working group
- establishing an effective secretariat
- constituting a small but effective team of policy analysts
- identifying a reasonable timetable
- defining key policy components early in the design phase
- promoting open and extensive consultations
- preparing a summary statement for cabinet approval

Key lessons from Benin's experience were that policy developers must not be overly optimistic about the time required to build consensus on a policy framework. Further, analytical methods are needed, as well as a vocabulary that promotes stakeholder participation in discussions. These elements help reconcile scientific and development interests and enable nonspecialists to appreciate the value of empirical evidence for decision making (Janssen, Perrault, and Houssou 1997).

Nigeria lies at the other end of the spectrum in terms of the degree to which an agricultural research policy has been made explicit: no formal policy exists in that country. The implicit nature of Nigeria's research policy has led to a highly unstable environment for agricultural research, particularly with respect to the level and consistency of government funding. Radical paradigm shifts are common, as research organizations experiment with ad hoc programs and institutional arrangements. Unsurprisingly, efforts to counter instability by diversifying sources of financial support for agricultural research beyond the public sector have yet to bear fruit (Idachaba 1998).

In Tanzania, agricultural research leaders have had limited success in convincing national policymakers of the link between agricultural technology development and dissemination and a country's ability to achieve development goals. This problem is not unique to agricultural research. Rather, it is symptomatic of a wider constraint, namely the absence of an enabling infrastructure (or framework) for science and technology. In the absence of a supportive framework, low perceived and actual returns to agricultural research become self-fulfilling prophecies (ISNAR 1997a).

Agricultural research policy in Peru is at a crucial stage in a multi-year reform process. A proposal currently under review outlines a system organized around three broad segments: a national agricultural technology council, a national institute for agricultural technology, and several regional technology centers. Fundamental to the proposed new structure is an emphasis on demand-driven agricultural technology development. Also important is a rational division of labor between public and private participants with respect not only to financing agricultural research, but also to providing extension services. Implementation, however, is recognized as the principal constraint. Major emphasis is thus being placed on drafting and enforcing a conducive legal setting and on involving key members of the political establishment in the reform process (ISNAR 1996).

The Netherlands has transformed its agricultural research system profoundly over the past 25 years, from a system dominated by central planning and control to one reliant on private initiative in a market-driven environment. One important lesson from this experience is the length of time needed to effect far-reaching policy changes. The initial proposal to privatize key services was made in 1986, but implementation did not occur for another decade. Even then, full institutionalization of the reforms came only after numerous complementary and facilitating adjustments had been made in the structure of the agricultural system, in the pace and direction of advancements in science, and in overall political and social ideology (Roseboom and Rutten 1998).

Several countries in west Asia and north Africa have a long tradition of agricultural research. Agricultural research institutes are often well funded and staffed. Even so, however, agriculture is often considered of secondary importance in national development and thus faces a constant challenge to maintain public support, particularly in the context of ongoing economic liberalization and structural adjustment. A major task for agricultural research policy in these countries is to articulate how technical innovation in agriculture can help facilitate and accelerate the transition from distorted macroeconomic policy regimes to less distortion-ridden, vibrant situations (Tabor and Janssen 1996).

Agricultural research policies in the newly independent countries of the Caucasus and Central Asia – as in all parts of the former Soviet Union – are developing in extremely unsettled conditions. The Soviet Union had established a highly developed agricultural research system, with the Academy of Agricultural Sciences as its central, most visible component. Although well defined and coherent, this system was not particularly efficient. Institutes and research organizations were often established to serve broad regional needs. Thus they were not necessarily relevant to local needs in the places where they were located. Since 1992 the new countries of the region have found themselves faced for the first time with the task of defining and developing truly national agricultural research systems and policies. Given that each inherited a particular set of organizations with their historical origins and linkages, it is clear that not all will prove useful for national purposes. In addition, these countries face the challenge of transforming all sectors and policies in the national framework in which agricultural research policy will fit (Morgunov and Zuidema 1999).

References

Elliott, H. 1996. Removing constraints in ACP agriculture: The role of policy and institutions. In *Priority Information Themes for ACP Agriculture: Proceedings of a CTA Seminar Held at Wageningen, The Netherlands, 30 September – 4 October 1996.* Wageningen: Technical Centre for Agricultural and Rural Cooperation.

Idachaba, F. 1998. Instability of National Agricultural Research Systems in Sub-Saharan Africa: Lessons from Nigeria. Research Report No. 13. The Hague: International Service for National Agricultural Research.

Idachaba, F. 1996. Agricultural Research Policy and NARS Instability in Sub-Saharan Africa. Mimeo. The Hague: International Service for National Agricultural Research.

Janssen, W., P. Perrault, and M. Houssou. 1997. Developing an Integrated Agricultural Research Policy: Experiences from Benin. Briefing Paper No. 33. The Hague: International Service for National Agricultural Research.

ISNAR. 1990. Organization and Structure of NARS: Selected Papers from the 1989 International Agricultural Research Management Workshop. The Hague: International Service for National Agricultural Research.

ISNAR. 1996. Peru's ministry of agriculture moves to restructure its agricultural technology system. *ISNAR Newsletter*, 31, August–December: 4.

ISNAR. 1997a. Policy and Financing of Agricultural Research in Tanzania: Proceedings of a First Workshop, 2–3 September 1996 and Follow-up Workshop 8–9 January 1997, Dar-es-Salaam, Tanzania. The Hague: International Service for National Agricultural Research.

ISNAR. 1997b. Strengthening the Role of Universities in the National Agricultural Research Systems (NARS) in Sub-Saharan Africa: Conceptual Framework. The Hague: International Service for National Agricultural Research.

Morgunov, A. and L. Zuidema. 1999. The Legacy of the Soviet Agricultural Research System for the Republics of Central Asia and the Caucasus. Discussion Paper No. 99-1. The Hague: International Service for National Agricultural Research.

Roseboom, J. and H. Rutten. 1998. The transformation of the Dutch agricultural research system: an unfinished agenda. *World Development*, 26 (6): 1113–1126.

Tabor, S. and W. Janssen. 1996. Structural adjustment, trade liberalization, and agricultural research policy in developing Mediterranean countries. Paper presented at the International Association of Agricultural Economist's 1996 inter-conference symposium on GATT implementation and structural adjustment in the Mediterranean Region, Rabat, Morocco, June 24–26, 1996.

Chapter 7
Science and Technology Foresight

Hans M. Rutten

Foresight uses a long-term view of the future of science and technology in the global economy and society. An interactive process, it seeks to identify and support viable strategies and actions by stakeholders. It is a relatively recent approach, distinguishable from predictive and forecasting.

What is science and technology foresight?

Science and technology foresight (STF) is the interactive process of systematically exploring future dynamics of science, technology, the economy, and society in order to identify and support viable strategies and actions for stakeholders (adapted from Martin 1995). It is a means of sustaining or increasing the contribution of science and technology to long-term social innovation. STF can take many shapes: a study (a research project), a series of debates, or perhaps a number of reports. Its four key characteristics are that it is process oriented, forward looking, science and technology based, and strategic.

An interactive process, STF involves a wide variety of players, experts, and change agents. This requires creation of (ad hoc) networks and the availability of open platforms for discussion. STF's forward view aims more towards exploring future dynamics than at predicting or forecasting them. Emphasis is on conceivable changes, rather than those most probable, in order to identify a broad range of opportunities and threats. STF is based on the premise that there is no single, most likely future. Instead, many possible futures are foreseeable, of which the most desirable is to be pursued. There is neither a pure science/technology "push" nor a pure demand "pull." STF combines both approaches. The philosophy is that of the "ambition-driven" strategy, in which the need for sustainable development to a large extent connects diverse stakeholders' ambitions. These ambitions serve as stepping stones towards strategies for exploiting new challenges and perspectives.

Yet one element still needs to be added to the definition: although most STF exercises are heavily or entirely focused on *areas* of research (e.g., biotechnology, productivity, food security), STF can also be a tool for identifying choices about the infrastructure of science and technology (e.g., suggesting organizational changes that can help society reap potential benefits of a new area of research). And, because of STF's two-sided approach (integration of science push and demand pull), it can enhance the effectiveness of science and technology while making society more aware of and sensitive to the work of research organizations.

In short, STF is a supporting instrument for priority setting, coordination, and innovation in science and technology. If substantial doubt or uncertainty exists about the present and future performance of an organization (be it a research organization or a scientific system), STF can serve as an all-encompassing instrument for strategic guidance.

Before the 1980s, STF was virtually absent or largely restricted to forecasting developments within science and technology; little attention was paid to socio-economic trends and needs. Also, technology fore*casting* was used to delve into the future, relying heavily on probabilities. Forecasting focuses mainly on the "most promising" and the "most likely" scientific and technological scenarios. STF has now developed into a more or less separate domain of foresight exercises. Irvine and Martin, from the University of Sussex in the United Kingdom, have played a crucial role in establishing this domain through their international studies and reports on STF and their roles as advisors in many parts of the world.

The fact that the foresight approach has gained ground, especially in recent years, is attributable to several factors. One is that many fore*casts* proved wrong. Another is the political, economic, and technological turbulence that became manifest in the 1980s, for example, the rise of newly industrializing developing countries, the breakdown of the socialist economies in Eastern Europe, and fast developments in microelectronics and biotechnology.

As the drawbacks of fore*casting* became obvious, governments felt the need to both rethink and strengthen the role of science and technology in society. By then, the STF approach as documented by Irvine and Martin (1984, 1989) and others seemed a logical tool. Particularly in the United Kingdom, France, Germany, the Netherlands, Hungary, Japan, India, Australia, New Zealand, Canada, and the United States, more or less nationwide STF exercises took place or were instituted as an ongoing element of the national science and technology policy framework (Irvine and Martin 1989, OECD 1996). In 1998, the Asian Pacific Economic Community (APEC) established a Centre for Technology Foresight hosted by the National Science and Technology Development Agency of Thailand. Alongside, and often as part of, these high-level, nationwide STF exercises, numerous individual science and technology institutes or individual economic sectors performed STF projects for their own benefit.

Over time and across countries, a wide variety of types of STF has evolved. A distinction already mentioned is that between nationwide and sector- or technology-specific exercises, as well as projects emphasizing future development in technology next to those with a more integrated view on science, technology, society, and markets. Martin (1995) presents a detailed typology of STF exercises. Interestingly, over the past decade, foresight exercises have come to rely less on expert opinions, in favor of wider participation by a multitude of stakeholders.

Doing science and technology foresight

Stakeholders and experts

STF exercises can be performed by individual research organizations or units within them, by research councils and science and technology departments in government, or by any other group who has a direct or indirect stake in science and technology. Regardless of who initiates an STF project, all these stakeholders must somehow be involved. For example, an STF performed by or on behalf of an agricultural research organization should involve scientists, managers, financiers and sponsors, policymakers, farmers, entrepreneurs from agroindustries and agrotrade, and, not least, representatives from societal groups (e.g., environmentalists and consumers).

In the early stages of the process, as soon as the objectives of the STF are clear and agreed upon, time must be spent inventorying decision makers, not only within government but also within research agencies and agricultural and agribusiness organizations, with regard to the objects of the exercise. Further, those affected by the current state of affairs and by future changes as "foreseen" by the STF must be drawn in. For example, a foresight exercise on the future role of science and technology in animal health should include representatives from farmers' organizations, government, veterinarians, veterinary scientists, animal husbandry researchers, stable constructors, policymakers, consumers, and perhaps others.

As to the experts involved, a segment of the stakeholder circle can and should be used, as well as those with "outside knowledge." That is, those who are expert in a more or less related field, but who are not part of the circle of "full experts" on the specific subject. Examples are experts on human health who also have some knowledge of animal health, experts on consumer concerns who are familiar with animal husbandry, and experts on quality management who have some knowledge of animal husbandry and food processing.

For all the stakeholders and experts involved, one crucial selection criterion should prevail: they must have the capacity (and the freedom!) to take a long view; that is, they should not focus too much on the present state of affairs.

A model of the process

STF is not a study project. It is a venture in which a wide range of information resources and (actual and potential) stakeholders are brought together to craft a more or less shared vision on the future role of science and technology. Already this implies that there can be no "cookbook" for how to do an STF project. Nevertheless, based on work by Irvine and Martin (1995) and the author's experiences with STF at the Netherlands Council for Agricultural Research, a rough framework can be constructed. It should be emphasized, however, that this framework is a stylized version of how the process can take place in reality.

Figure 1 presents the process of foresight in science and technology in general. It can also be used when the STF focus is narrowed to, for example, agricultural research or another specific domain of science and technology. The figure (read from the bottom upwards) identifies the most significant stages of the STF process and the requirements and actions at each stage. A stage is loosely defined according to the main task at hand. Attention is also given to the by-products or intermediate results of the process, an aspect often ignored but crucial to the impact and success of STF exercises. Those responsible for implementing STF are advised to value and monitor these by-products carefully. Certainly they should resist the temptation of considering the final report to be the only relevant product of the effort.

Four stages outlined in figure 1 deserve separate consideration:

- defining of the core objective(s)
- selecting participants
- carrying out the strategic analyses
- converging to promising strategies and related actions

Defining the core objective(s)

Let us assume that the minister of agriculture of country X has decided to undertake a foresight project for agricultural research. The minister assigns an independent group to carry out the project. Yet how should the project's focus be determined? There are three options:

1. determine the most challenging fields of agricultural science and technology, for example, the future of soil sciences, agricultural biotechnology, veterinary research, or social sciences
2. identify the most pressing societal issues related to science and technology, such as food security, environmental quality, or consumer concerns
3. determine future strategic options for the most promising or otherwise important production branches, for example, aquaculture, forestry, or animal husbandry

In many STF projects, the first type of objectives (fields of science and technology) dominate, with the result that the project soon becomes a technology-push one. However, stakeholders outside the circle of scientists may tend to lose interest in such a process. If the second type of objective is chosen as the starting

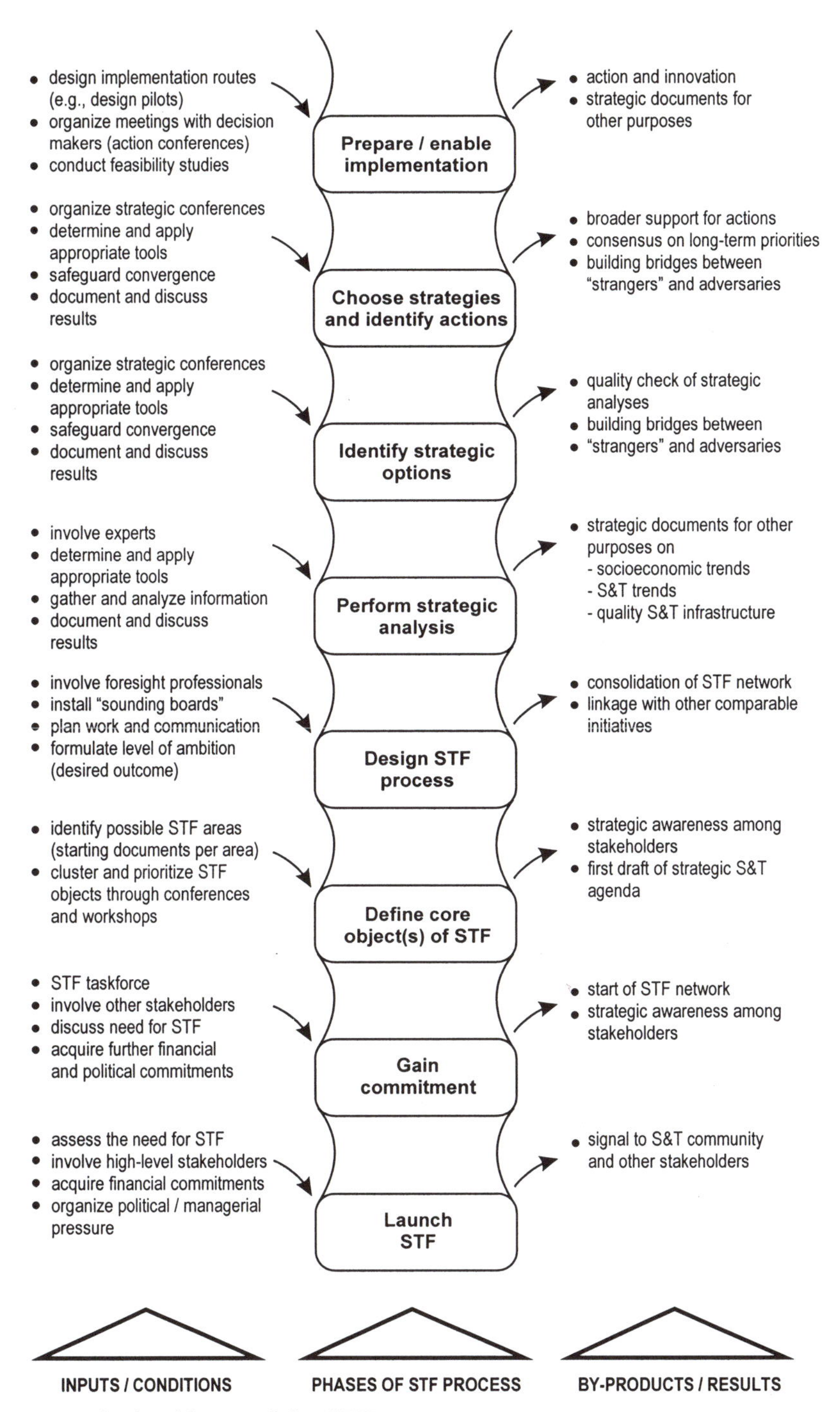

Figure 1. The backbone of the STF process

point, the project may become a program for social change, with insufficient attention given to the innovative potentials of scientific and technological developments. Scientists and the like might be unwilling to invest their energy in such a venture. The third type of objective is less vulnerable to the risks associated with the first two. In addition, it offers opportunities for integrating future scientific and technological opportunities with future societal needs. The disadvantage however, is that the smaller the production branch chosen, the narrower the scope of the foresight exercise will be.

Selecting participants

In designing the process (i.e., deciding who does what, when, how, and with whom) the involvement of multiple participants is carefully planned. After all, not all potential stakeholders or experts can actively participate at all times. A useful technique is to work with four circles of participants (see figure 2).

First a small group of *high-level stakeholders* should be drawn in. These participants should have demonstrated commitment to and responsibility for the entire process and its outcomes. The group should include top policymakers and opinion leaders. In order to be effective, the foresight exercise needs their commitment to the processes that will be set in motion, as well as to practical outcomes. This requires their active involvement in the beginning and at the end of the process.

Second, depending on the number of specific objectives chosen as focus points, one or more *sounding boards* should be installed. These boards consist primarily of a mixture of stakeholders and experts whose main role is to reflect upon the process and assist the project team in making decisions about how to proceed.

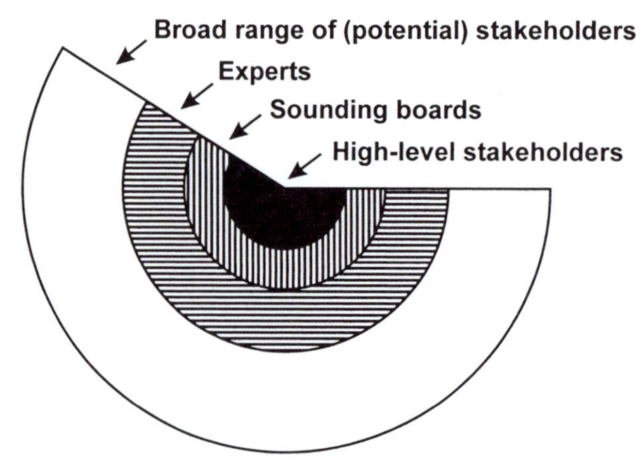

Figure 2. Circles of participants

Third, for specific tasks during the project, *experts* should be invited. Their duties may vary from facilitating group meetings, to writing provocative essays and conducting in-depth analyses.

Fourth, for workshops and, more so, for strategic conferences, the widest circle of participants is needed, consisting of stakeholders from research organizations, academia, business, pressure groups, government, and others. None of these groups should dominate, however, at least not quantitatively. They should be selected based on their willingness and ability to look beyond today's problems and search for common aspirations. The involvement of stakeholders from pressure groups and the business sector requires special attention, not only because of the seemingly natural antagonism that often exists between the two, but more importantly, because of these groups' relatively greater distance from the domain of science and technology. Still, their participation is crucial, as they are representatives of the main audience that is to be addressed.

These four circles of participants and the way each circle is deployed constitute the organizational backbone of the project: without planned interaction between a variety of interests and perspectives, the strategic intention of STF can not be realized.

Strategic analyses

The analytical part of an STF project broadly consists of a "SWOT"-type analysis, which looks at the **s**trengths and **w**eaknesses of the actors involved and the **o**pportunities and **t**hreats that might come along with future developments. Based on Martin (1995), the main inputs required for this analysis are assessments of four areas:

1. evolving societal needs and opportunities
2. emerging scientific and technological opportunities
3. the current socioeconomic situation, including capacity to exploit new scientific and technological opportunities
4. the current situation of the science and technology community, including its capacity to relate to potential beneficiaries in society

Thus, two SWOT studies are needed, one focusing on the current situation and the other examining potentials. The results are used in the strategy and action-generating process (figure 3).

A large number of disciplines (e.g., economics, political science, natural sciences) and research methods (both quantitative and qualitative) can be used at this stage of the project. Here, much can be learned from analyses done in foresight exercises in other branches or countries.

From a cognitive point of view, thorough, in-depth SWOT studies help establish a solid foundation for development of science and technology. There are,

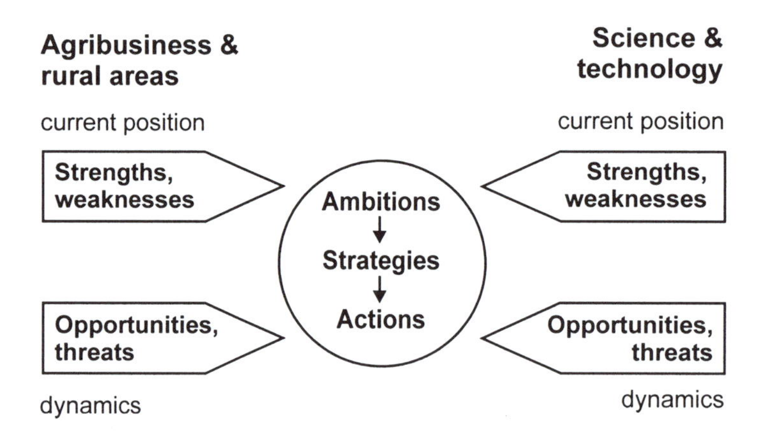

Figure 3. The main elements of strategic analyses

however, two dangers in putting a large share of the resources available to the project into the SWOT study. One is that because of the often time-consuming character of these studies, the project may lose momentum. Another is that the research approach to SWOT may become biased towards facts and figures. A proper SWOT focuses on perceptions and aspirations, as well as facts and figures. Yet to formulate a strategy, it is at least as important to have objective information on the status quo and possible future directions, as to have information on how various stakeholders perceive the present situation and what their long-term aspirations are. To avoid these dangers the results of strategic analyses should be treated primarily as a communication tool in the strategy- and action-generating process, rather than as the analytical input to a process of rational planning.

Converging to promising strategies and actions

The proof of the pudding is in the eating: when a STF project does not lead to a shared vision on "where to go" or a program of concrete proposals on "how to get there," the project has failed to deliver. Again, there is no magic formula for success. An important lesson learned in the foresight projects of the Dutch National Council for Agricultural Research (NRLO) is that it is helpful to start a STF project by putting a lot of energy into considering the kinds of strategies and proposals for action that might result from it. This "thinking ahead," if pursued consistently throughout the project, may considerably improve its effectiveness.

Tools

We have called STF an *approach* to indicate that it is not a clear-cut methodological tool. Instead, it uses a broad range of more or less established tools, such as Delphi surveys, statistical trend analyses, and other quantitative or qualitative forecasting techniques. All are not necessarily part of every STF project, however. The same is true for tools that usually build upon forecasting techniques, like cross-impact analysis (in which the cross impact of future developments is analyzed), strategic SWOT analyses, and scenario planning (see Johnson and Paez, this volume). Often, very down-to-earth communicative tools, such as in-depth interviews with a broad range of stakeholders, or provocative essays by a small number of experts turn out to be at least as effective as the more analytical tools.

Thus, the toolbox available to those responsible for the STF process contains many items. Which of these items are used depends not only on the financial and professional means available, but also on the time available for the project, and, most importantly, on goals. If the goal is primarily an (ex ante) evaluation of current strategies for science and technology, then analytical and quantitative tools are most appropriate. If the goal is primarily to *generate* strategies, then more qualitative and communicative tools are appropriate.

Finally, two often undervalued managerial tools should be mentioned. One is "knowledge management," that is, the systematic approach to the ways in which relevant knowledge is being created and communicated. The second is "client management." Since the quality of the foresight exercise hinges on the value participants attach to it, all participants should be seen and treated as valued clients. This requires detailed knowledge of their preferences and competencies as well as, whenever possible, personalized communication strategies.

Relevance for agricultural research

Based on STF exercises in the Netherlands and other countries, some observations can be made on STF's usefulness in agricultural research planning:

- STF is a powerful tool for breaking down or, at least, reducing the barriers that often exist between various parts of a country's science and technology community, not only barriers separating researchers, policymakers, and agribusiness, but also those between science and technology disciplines.
- STF helps guide ongoing structural transformations within agricultural research systems.
- As a result of the often intensive participation of agricultural scientists, STF helps improve the strategic orientation of individual research organizations.

Pitfalls and success factors

As Martin (1995, 1996) writes, no individual foresight approach is perfect. Each has its own strengths and weaknesses, and foresight rarely works well on first attempts. So what contributes to the effectiveness of STF? One way to answer this query is by pointing out three of the most serious pitfalls of the approach.

First, STF hinges on intensive interaction with a wide variety of stakeholders. On one hand, this extensive participation increases effectiveness because of the (broad) commitment that may result and because many stakeholders already have a number of future perspectives in mind and already follow future-oriented strategies. On the other hand, effectiveness may be jeopardized if insufficient organizational effort is put into maintaining and building on broad participation. Note that although a team of salaried professionals is required, the largest part of the effort put into a STF exercise must come from stakeholder volunteers. The larger the number of volunteers, the more critical is the task of keeping the process interesting and productive for them.

Second is the challenge of striking a balance between ambitious, global strategies for science and technology and remaining practical and specific. Too often, strategies do not work because those involved find it difficult or impossible to translate the path(s) laid out in the strategy into actions that make sense here and now. Perhaps this phenomenon explains the sustained popularity of classical strategic *planning*, since it simply prescribes what those involved need to do. The purpose behind STF, however, is not to prescribe strategy-related actions, but to inspire those involved to reflect on the long-term consequences and possibilities of decisions they make today. In order to remain effective, the process requires mechanisms to balance stakeholders' present worries and hopes with future opportunities and threats.

A third pitfall is when the circle of people involved becomes too "familiar." STF is not about science and technology per se, but about the long-term contribution of science and technology to an organization, firm, industry, or society. This means future possibilities created by scientific and technological progress must be integrated with future needs and opportunities facing society (or the organization, firm, or industry). Experience shows that stakeholders from research organizations and from educational institutions tend to focus too much on the potentials of science and technology and be too defensive of past achievements, whereas *other* stakeholders tend to be too focused on the immediate usefulness of science and technology to solve present-day problems. To balance these two tendencies, outside experts (who have no immediate stake in the science and technology domain involved) should be involved in STF as well as innovative and strategically thinking representatives from outside the agricultural sector, as they can contribute new perspectives to the problems perceived.

Apart from these pitfalls to be avoided, a number of straightforward success factors can be given:

- Ensure that the STF is commissioned by a private or public body that can act as a (nonexclusive!) client or customer.
- Provide feedback to participants and communicate with potential participants. If participants lose contact as a result of slow or no feedback from the project team, the success of the project is at risk. Through various means of communication (newsletters, leaflets, reports, small-scale presentations) participants can be triggered to again reflect on intermediate project results.
- Include a wide range of skills and expertise on the project team.
- Ensure that the project team is well informed about experiences with innovation processes in other fields (e.g., industrial companies and the services sector), because essentially the STF venture is about social innovation and how science and technology might best contribute to achieving this goal.

Examples

There are few examples of targeted, agriculturally oriented STF exercises. In a nationwide STF in the United Kingdom starting in 1994, two of the 16 so-called "foresight panels" that were established had more or less direct relevance for agricultural research: the panel on agriculture, natural resources and the environment (later divided into a panel on agriculture, horticulture, and forestry and another on natural resources and the environment) and the panel on food and drink. Later, in 1998 an additional cross-panel, the "food chain group," was formed.

An STF fully oriented towards agriculture was conducted by the NRLO. In 1994 the Dutch Ministry of Agriculture, Nature Management and Fisheries asked NRLO to conduct a series of foresight exercises for agriculture, agribusiness, and rural areas. The assignment started in 1995 and was concluded and evaluated in 1998. To do the project, NRLO was restructured into an independently operating, small organization of professionals with a supervisory board and the financial means and mandate to build a large network of stakeholders and experts.

A few highlights from the NRLO program show how a targeted STF can be implemented. The program began with a conference in which a broad group of stakeholders in agricultural science and technology gave their opinion about themes thought to be key for the next 15 years. This resulted in a long list of themes that was subsequently narrowed to five clusters:

1. redefining the role of primary agriculture in society
2. rural development, that is, establishing sustainable interaction between the numerous and sometimes conflicting functional claims on access to and use of rural resources
3. sustainable development of agribusiness in view of changes in consumer demands (individualization of preferences, as well as concerns over animal

welfare), environmental standards, and international markets (globalization and liberalization)

4. sustainable development of fisheries and aquaculture in view of increasing political problems involving fisheries stocks and increasing worldwide demand for animal protein
5. organization of innovation processes in view of structural changes within agriculture and changes in the funding and steering of research, education, and extension.

As these themes represented the outlines of the demand-pull element of the STF, early on it was decided that the program should also cover foresight of more or less autonomous developments that were taking place within domains of science and technology that may be relevant to the above themes. Sensor technology and molecular biology are examples of these domains. At a later stage a third element was added to each domain: the assessment of strengths and weaknesses of the present agricultural research system.

The following data and experiences give an impression of the program's results:

- Much effort has been put into the dissemination of intermediary documents. More than 100 NRLO background studies and 13 newsletters describing and discussing the program were produced.
- Some 30 workshops were organized involving between 600 and 700 stakeholders and experts.
- For each of the five themes, integrated reports were drawn up called the "foresight reports." In total, 10 foresight reports were published.
- Stakeholders increasingly explicitly mentioned activities and actions that fell under the program as a source of inspiration for their own operations.

In conclusion, there is no one best way to perform STF. Nor is STF the only way to priority setting, coordination, and innovation in science and technology. Experience shows, however, that STF can be a powerful tool, provided the organizational conditions are conducive and those involved remain enthusiastic about building their own future and sharing hopes, fears, and potentials.

References

ASTEC. 1996. Developing Long-term Strategies for Science and Technology in Australia. Canberra: Australian Science and Technology Council.

Cabello, C., F. Scapalo, P. Sørup, and M. Weber. 1996. Foresight and innovation: the role of initiatives at European level. *The IPTS Report,* Nr. 7 (September).

Irvine, J. and B. R. Martin. 1984. Foresight in Science: Picking the Winners. London: Pinter Publishers.

Irvine, J. and B. R. Martin. 1989. Research Foresight: Creating the Future. Zoetermeer: Netherlands Ministry of Education and Science.
Martin, B. R. 1995. Foresight in science and technology. *Technology Analysis and Strategic Management*, 7 (2): 139–168.
Martin, B. (ed.). 1996. Technology Foresight in Europe: Results and Perspectives. Report of a TSER/ETAN Workshop, Brussels, April 1996.
Rutten, H. 1997. Coping with turbulence: strategies for agricultural research institutes. NRLO Report 97/27. The Hague: Dutch National Council for Agricultural Research.
OECD. 1996. Special Issue on Government Technology Foresight Exercises. STI Review No. 17. Paris: Organisation for Economic Co-operation and Development.

Chapter 8
Strategic Planning

Carlos Valverde

Strategic planning is a means of adjusting an organization's objectives, activities, and management of resources in response to changes in its external environment and client needs. For agricultural research organizations, strategic planning positions the organization within the context of national development plans, enables effective and efficient use of scarce resources, and identifies structural changes needed for good performance. Strategic planning often serves as the foundation on which other types of agricultural research plans rest.

What is strategic planning?

Strategic planning is a process by which a future vision is developed for an organization, taking into account its political and legal circumstances, its strengths and weaknesses, and the threats and opportunities facing it. A corporate strategy is the end product of strategic planning. It articulates the organization's "sense of mission" and maps out future directions to be taken, given the organization's current state and resources. In general, strategic planning seeks to clarify how the organization should deal effectively and efficiently with

- its external environment
- internal resources and capabilities
- actions necessary to carry out change
- how actions should link together as the strategy unfolds

Organizations use strategic plans primarily to adapt to a competitive and dynamic world. Strategic plans also communicate organizational objectives and domain legitimacy (its right to undertake the designated activities and tasks). In simple terms, strategic planning engages the organization in determining what it is, what it wants to be, and how it anticipates getting there. It is also a process that develops ownership and understanding of the organization and engages various actors in decision making in a transparent fashion.

Different approaches can be used in doing strategic planning; there is no universal or best practice. Further, it is important to remember that changes in the environment occur continually. Strategies, therefore, need to be updated from time to time. While having a strategy document as a record of decisions made is useful, it must be a "living document" that can be adapted to changing needs and circumstances of the organization and its clients.

Doing strategic planning

The aim of strategic planning is to establish a flexible link between the management of the organization's internal resources and its interactions and relationships with producers, agroprocessors, government, and external actors, all within the existing economic, social, and institutional environment.

Strategic planning should be organized in light of organizational capacities, time restrictions, and resource constraints. Effective leadership for the process is essential. Establishing a functionally responsive unit or ad hoc committee is usually one of the first tasks of a strategic planning committee appointed by senior management. The committee interacts with the organization's top management and staff and is charged with designing the planning process and, eventually, compiling and drafting the final strategy document. The composition of the strategy committee depends on the scale, scope, and circumstances of the organization. For example, it could comprise different levels of management (e.g., national and experiment station directors, program leaders, and project managers), assisted by specialized staff from within the organization or elsewhere. The committee needs direct interaction with key decision makers in the organization, to discuss the logic of the strategic planning process and to reach agreement on the steps to be followed.

To be successful, strategic planning must be visibly supported from the start by the organization's highest management echelons. If the research establishment falls under a higher authority, for example, a ministry of agriculture, the strategic planning exercise must be politically supported by the relevant minister. At the same time, dominance of the process and decision making by the organization's leaders must be avoided. Rumors and gossip, where they arise, should be turned into opportunities for open debate, else they prove destructive. The best remedy for management dominance and rumors is to communicate results of ongoing activities as soon as possible. Strategic planning activities should involve interaction and participation of those affected by its results. Adequate time should be allowed for the process, to develop a well thought out and comprehensive strategy and its related documents. Reasonable contributions from staff and other stakeholders should be taken into consideration.

Most approaches to strategic planning follow a sequence of steps. The steps, their content, and the manner of their implementation vary considerably, however. Strategic planning may be incremental in nature: the pathway for institutional development may be defined by exploring future trends and defining the gaps to be filled in order to respond to changes. Or it may lead to renewal of the organization, with future strategic issues identified. The strategic plan focuses on how the organization should reposition itself to address new conditions and issues adequately. But, "[a] key point to be emphasized again and again is that it is strategic *thinking and acting* that are important, not strategic planning. Indeed, if any particular approach to strategic planning gets in the way of strategic thought and action, that planning approach should be scrapped!" (Bryson 1995: 2).

Rather than presenting one approach to strategic planning, figure 1 illustrates four basic building blocks used in most strategic planning processes. These are not steps in the process, but groups of activities, each of a different nature. The first block is *analytical.* It focuses on identification of trends in the organization's external environment and an internal analysis of mandates and programs. The second block is *normative* in nature. It results in choosing a future strategic direction for the organization. Specific activities here may include identification of strategic issues and objectives and the development of a mission statement. The third block identifies *actions* required to achieve the objectives. It includes activities such as gap analysis, constraints analysis, and priority setting. The fourth block comprises *planning for implementation* activities. It includes the design of feedback mechanisms (monitoring), structures required for implementation, resources needed (staff, funds), responsibilities, and leadership issues.

Analyzing context and trends

One of the first steps in strategic planning is examining the organization's external environment. Threats and opportunities, for example, an unstable funding base or dramatic political changes, are identified and analyzed to ensure the organization's good performance and even survival in a competitive, changing environment. This procedure is known as the "SWOT analysis" (analysis of **s**trengths, **w**eaknesses, **o**pportunities, and **t**hreats in the environment). In the

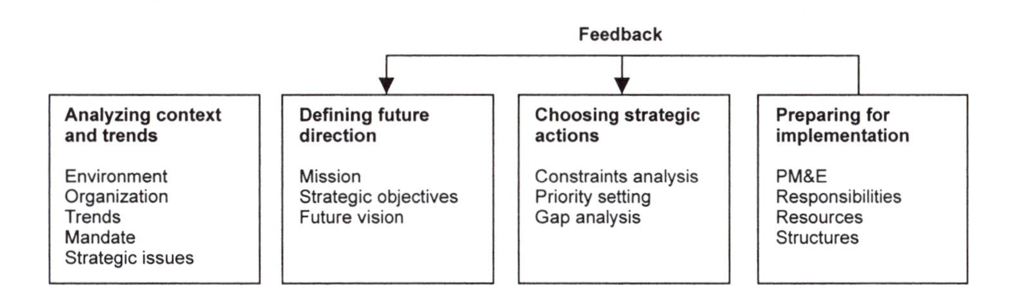

Figure 1. Four basic building blocks of strategic planning

case of public-sector agricultural organizations, national development policies should guide the strategic planning effort to ensure complementarity with government objectives and trends in the sector. Another crucial element of the external environment analysis is identification of the roles and responsibilities of other actors (partners and competitors) to maximize coordination of activities. Any disagreement among actors or inconsistencies in policies should be identified and resolved through discussion and consensus.

The assessment of the internal environment focuses specifically on the organization. Its current mandate, objectives, mission, strategies, priorities, achievements, and resources are examined in terms of their strengths and weaknesses. This process involves a careful and extensive collection of information that will be useful throughout the strategy's development and implementation.

The scale and scope of research to be carried out are influenced by technical capacity, the nature of relationships and linkages with other actors, and new technological trends and demands. Therefore, such factors should guide the analysis of the external and internal environment. For instance, the analysis should consider whether research will be basic, strategic, applied, and/or adaptive given the organization's capacity and mandate in the research and development system. This phase of strategic planning provides information for decisions on organizing and complementing efforts among actors (public, private) and partners (national, international), and also facilitates subsequent definition of the organization's (new) strategy.

The analysis can be supplemented by identifying key or central strategic issues that the organization should address. These issues are the critical challenges that the organization will respond to in light of clients' needs, threats and opportunities in the environment, and stakeholders' interests. Issues are strategic only if there are important consequences for not addressing them; the strengths and weaknesses of the organization should be considered in identifying the issues. Once the strategic issues are defined (often in special stakeholder working session), the strategy committee can identify potential actions (i.e., strategies) to address them.

Defining future directions

Research organizations engage in strategic planning for a number of very different reasons. Strategic planning may be routine (e.g., a five-yearly event). It may be prompted by financial crisis; or the organization may face changes in the medium term that force it to rethink its strategic options (e.g., if a research organization is being privatized).

In any case, strategic planning work should start by taking a long-term perspective and envisaging what the organization should look like in, say, five or 10 years. It is equally important to ensure that staff share the vision. That means the organization's management and the strategy committee must have a good knowledge of the major environmental forces, technology and knowledge demands, and other strategic issues. Before deciding on a strategic direction for the organization, two elements should be considered: the needs and conditions of the main clients and the organization's mandate and mission.

Priority clients and their needs are crucial considerations. Farmers and their organizations, processors, input providers, and government agencies are among the main client groups of most agricultural research organizations. An up-to-date assessment of client needs may be conducted if such information is not already available. Such an assessment may include special workshops and surveys to involve users of the organization's scientific knowledge and its products and services (e.g., extension services and farmers' organizations). The type of producer being targeted should also be determined (e.g., large-, medium-, or small-scale producers) as well as the conditions under which they work.

Changing environmental factors may necessitate adjustments to an organization's mandate or mission. An organization's mandate specifies its role, responsibilities, and authority. The organization's mission is a precise statement of its reason for existence. It broadly characterizes the targeted clients, objectives, and activities. It is important to understand that while the direct clients of agricultural research are often farmers or government agencies, benefits of research may be spread more widely throughout society, also reaching consumers. A stakeholder analysis helps define the organization's mission in relation to clients' needs.

The aim of top management and the strategy committee should be to challenge the knowledge, expertise, and creativity of staff, so they can articulate how the organization can best contribute to national development. A mission statement is one mechanism that can help achieve this. Its development is facilitated by posing fundamental questions: "What does the organization contribute to society and to its stakeholders?" "What will the organization be like in five or 10 years?" "Who will be its main stakeholders?" "What will be its primary research and service outputs?" "What types of staff will it have and what will be their qualifications?" And, "What will be its culture and shared values?"

The mission statement represents a commitment by management and staff to a longer term organizational goal or preferred state. Based on a future vision, a mission statement can be useful for identifying and, in some cases, justifying future areas of research. Issues of trust, respect, dignity, commitment, integrity, and accountability are often reflected in a "vision statement." Such values are important because they affect strategic decisions and objectives. Furthermore, explicit values help determine an organization's image within and outside its immediate environment.

Analysis of the external and internal environment and the identification of mandate, mission, and vision serve as the foundation for the next major step in strategic planning: analysis of the current versus the preferred future state of the organization.

Choosing strategic actions and approaches

The reason for analyzing the current versus the future state is to compare the organization's actual state of affairs with its preferred future situation in order to identify strategic changes that need to be made. The analysis must provide a clear picture of the changes required to progress towards the strategic objectives identified, and it should lay the foundation for the change process. The comparison encompasses the organization's mandate, research priorities, objectives, products, and services both in their current and their desired future forms.

A variety of tools can be used to facilitate the step, including development of a matrix that specifies present and future roles of the programmatic units. Such a matrix may lead to the conclusion that changes are needed in the organization's structure or research program setup. The scale and scope of research activities are key determinants of organizational and program size and structure. Priority setting helps to define the relative importance of different programmatic units or research thrusts. It is used to select the most important and essential research to be undertaken and to provide defendable reasons for the decisions made. Priority setting also helps determine the level of financial and human resources needed for a specific period of time (see Contant, this volume). For purposes of strategic planning, it is not yet necessary to formulate the details of specific research programs or projects. But it is important to have an idea of their relative sizes.

Consideration should further be given to external and internal linkage mechanisms or strategic alliances required for the "new" organization to fulfill its commitments. This entails identifying the ways technology, information, and resources will be provided to internal units and external partners.

Preparing for implementation

This step determines how the organization will operate to produce, supply, and deliver research and service outputs to clients. It is what organizational management strategists often refer to as "the process of institutional change." The role of the various functions within the organization and the kind of structure and organizational model to be adopted need to be defined (or redefined). Functional analysis is the major tool for this step and serves to define roles within the organization as well as its structure. Six principal functional areas should be addressed:

- governance
- leadership
- planning, monitoring, and evaluation

- management of information and financial, human, and physical resources
- technology dissemination and transfer
- support services, including accounting and budgeting, purchasing and stores, maintenance, laboratory facilities, travel, computer services, biometry, and logistics

The results of functional analysis and the definition of programmatic research areas allow projections to be made of the financial investments required over the strategic plan period and the institutional changes needed to implement the plan. The strategic plan thus becomes a record of decisions and a framework for programming and investment in the organization.

Drafting the corporate strategy and gaining approval for it within the organization and among external stakeholders and clients is only part of the process. The strategy must also be translated into action, or implemented. Typically, planning for implementation involves designing action plans to ensure that strategic planning is a flexible, dynamic, continuous process sustained within the organization. Action plans may identify roles and responsibilities related to strategy implementation, as well as those accountable for specific activities. They may also specify sources and means of acquiring resources. Action plans, finally, provide the means to monitor the strategy and understand "what happens" to the plan when it is put into practice (see Horton and Dupleich, this volume). This information can be used to ensure that the strategic plan is periodically updated in relation to changes in the organization and its environment.

Relevance for agricultural research

Within the agricultural sector, public research activities that create or adapt technologies are under pressure to apply effective management practices. One such practice is strategic planning, which introduces new paradigms of thinking to the management of agricultural research. "Strategic management" tools such as strategic planning allow government institutions, including national agricultural research organizations to deliver efficient and effective products and services to their clients.

For agricultural research organizations, strategic planning serves a number of other functions as well:

1. orienting the organization with regard to its environment and a desired future state
2. taking into account the organization's strengths and weaknesses and the threats and opportunities facing it
3. determining goals and how to reach them
4. orienting the organization toward its clients' needs and demands

5. establishing various elements of direction and management, in particular, vision, mission, objectives, and structure and translating strategies into action plans
6. determining necessary changes and adjustments within the organization, including policy, administration, and management
7. incorporating experiences gained through a continual process of review, ensuring that the strategic plan reflects lessons learned and environmental changes

Finally, most national agricultural research managers and directors seek ways to use and distribute their scarce resources more efficiently, in order to contribute more effectively to agricultural development in their countries. For this reason, strategic thinking and management will continue to be a necessity in agricultural research management.

Examples

In general, public-sector agricultural research organizations' institutional, operational, and administrative frameworks are characterized by rapid political and policy changes and by strong socioeconomic pressures. Achieving consensus to guide strategic planning is therefore both difficult and rare. However, the underlying analysis and management's choices of actions to take in response to the analysis can contribute to building and maintaining support for and commitment to national agricultural research and its clients.

The following examples show how some countries and organizations have adapted strategic planning processes to their own capacities, needs, and constraints. Various approaches are incorporated.

Palestinian Authority

Following the endorsement and adoption of an agricultural research and extension policy in 1998, the Palestinian Ministry of Agriculture (MOA), in collaboration with ISNAR, began formulating a strategy for research and extension. This strategy followed on policy objectives that had been set in light of thorough analyses of external and internal threats and opportunities and the strengths and weaknesses of the Palestinian research and extension system. The strategic planning initiative constituted the Palestinian Authority's first systematic planning exercise for agricultural research and extension.

The strategic planning process incorporated analyses of the external environment for research and extension. Threats and opportunities were also identified, both for the MOA and other national actors, and future roles of other stakeholders in research and extension were suggested. The analysis addressed structure and organization; current responsibilities, activities, and outputs; resources;

and management processes. The strengths and weaknesses of organizations involved in agricultural research and extension were also identified.

Strategic issues pertaining to MOA research and extension, and strategies that address these issues, were then defined. These issues and strategies, identified separately for the Department of Extension, Information and Applied Research, and the National Agricultural Research Center, fell into three categories: institutional, socioeconomic, and those related to technology or production. Mission statements and contingency plans for the organizations were also developed.

Finally, the institutional changes needed to implement the strategy were identified. These addressed structure and organization and management processes, including planning, monitoring, and evaluation; information management; human resource management; and linkage planning.

A major difference between the approach used in the Palestinian Authority and the methods described in this chapter is that the Palestinians carried out program planning and priority setting after strategic planning was completed in a separate operational planning initiative.

Uruguay

One of the most extensive applications of strategic planning in Latin America in which ISNAR has been involved was in Uruguay. From this strategic planning exercise, a new decentralized body, the Instituto Uruguayo de Tecnologia Agropecuaria, was created.

In 1985, the Ministry of Agriculture and Fisheries appointed a task force to develop a proposal for reorganizing the country's system for technology generation and transfer and to improve planning and programming of agricultural research. Scenario diagnosis indicated that the agricultural research program was not sufficiently linked to national development plans and formal client participation was lacking. Critical areas identified included financial and human resources, underutilized facilities, and rigid and bureaucratic administrative norms.

Emphasis in the design stage was placed on the preferred features of the new organization and methodologies, mechanisms, and instruments to manage the critical areas noted above. Partnership with the Inter-American Institute for Cooperation on Agriculture (IICA) was also included in the technical design phase to comply with IICA's desire and willingness to collaborate with the Uruguayan government in agricultural research. IICA's assistance was subsequently directed to three areas: five-year research planning and programming, technology transfer, and financial administration. The strategy committee directed ISNAR to focus its assistance on agricultural research policy and priority setting, definition of the structure and organization of the new institute, and human resource planning.

While reinforcing the importance of an iterative and cumulative process of strategic planning, INIA's experience reflects the significant demands placed on national agricultural research organizations and the different types of agencies that can become involved in the process. Benefits can be accrued through strategic planning, as evidenced in the case of Uruguay: this process created a new and revitalized research system, capable of responding to dynamic conditions in the country (Valverde 1995).

The author thanks Govert Gijsbers, Helen Hambly Odame, Warren Peterson, and Michèle Wilks for their contributions to this chapter.

References

Bryson, J. M. 1995. Strategic Planning for Public and Nonprofit Organizations: A Guide to Strengthening and Sustaining Organizational Achievement. San Francisco: Jossey-Bass.

Bryson, J. M. and W. D. Roering. 1987. Applying private sector strategic planning in the public sector. *Journal of the American Planning Association,* 53 (1): 9–28.

Mintzberg, H. 1994. The Rise and Fall of Strategic Planning. Hemel Hempstead: Prentice Hall.

Ozgediz, S. 1987. A Strategic Planning Process Model. CGIAR Secretariat. Mimeo. Washington, D.C.

Valverde, C. 1995. Institution Building in NARS: The Case of Uruguay. The Hague: International Service for National Agricultural Research.

Chapter 9
Master Planning

Helen Hambly Odame

An agricultural research master plan is a long-term plan that deals with policy, strategy, research priorities, programs, and the resources needed to implement the plan. Master planning is a process; an official plan is its product. In practice the term master planning is used rather loosely to refer to a variety of different types of plans. Both the process (partners, context, and conditions) and the content (policies, research priorities, programs, and resources) vary considerably. Divergence among master plans also reflects the extent to which master planning is being used within agricultural research systems as an integrated mechanism for institutional development, as opposed to a "one-off" requirement for donor investment. A common element among master plans is their attempt to be comprehensive in addressing all relevant aspects. That is why master plans rely to a large extent on external inputs and resources.

What is master planning?

The concept of master planning originated in the fields of physical planning and civil engineering. It has been used particularly in the context of major infrastructure projects such as the development of new airports or universities. A master plan provides a blueprint for institutional development at the highest level. It is a long-term, overall plan that may incorporate a variety of more detailed operational plans.

Definitions of agricultural research plans may vary across regions, countries, and organizations. ISNAR has defined agricultural research master planning as follows:

> a process by which a national research system analyzes its present and future environment, defines its medium and long-term goals and objectives, and develops a plan based on priorities and available resources to attain these objectives and goals. An integral part of this process is the building of a sustainable research capacity which would enable the system to respond appropriately to the changing needs of the agricultural industry, and the nation. All these must be based on a clear and realistic vision of the future of research, the design of relevant and effec-

tive programs, and an assessment of the resources needed to achieve this vision (ISNAR 1989).

A master plan for agricultural research is a macro- or policy-level plan with a long-term perspective. A master plan is more detailed than a strategic plan and typically forecasts policies, priorities, and programs over a 10-year period with detailed recommendations on programs, projects, organization, structure, management, and resource requirements (financing, staff, and facilities) for an intermediate period. The detailed character of an agricultural research master plan and the significant effort that is invested in its preparation can potentially reactivate weak agricultural research organizations and support long-term institutional development.

Although the concept of master planning may be interpreted in different ways, plans do tend to share some common elements. First, master plans aim to be comprehensive. They try to incorporate all relevant elements in a single document in a pragmatic manner.

There is less theory on master planning than on strategic planning, even though the comprehensive approach of agricultural research master plans appeals to many countries and donors. This leads to an important second element of research master plans: they are often prepared in the context of donor-funded research support programs. As a result, financial and other resource subplans are included within the master plan, whereas strategic plans may not necessarily include these components. As the need to develop master plans is often externally motivated, they tend to follow a predetermined design rather than a process approach. Master plans are often prepared in a top-down manner, sometimes with heavy reliance on outside advisers. Intermediate agencies may be involved in assisting national agricultural research organizations in developing a master plan. In agricultural research, ISNAR and the Food and Agriculture Organization of the United Nations (FAO) are the two organizations that have been most involved in master planning.

Doing master planning

Current practice in master planning points to no generally applicable model. It is thus difficult to define a universally valid and acceptable approach to both the process and the content of a master plan. For this reason, it is often hard to distinguish between the content of a master plan and other plan types such as strategic plans. One study by ISNAR found that in sub-Saharan Africa, agricultural research master plans vary considerably in terms of their process and content. This variation is particularly evident in terms of the stakeholder consultation and the methodology behind priority setting. Plans also vary according to operational matters, such as the scale and scope of research programs and resource allocation (Hambly and Setshwaelo 1997). Despite the lack of a set model and method for master planning, a number rubrics of tasks are commonly involved in drafting the plans.

Research policy

- Ascertain, in consultation with the relevant stakeholders, the government's overall priorities for agricultural research in the country.
- Identify current and future policies with regard to funding and resource allocation.
- Recommend mechanisms for formulating and reviewing research policies.

Research strategies

- Determine realistic research goals, objectives, and approaches.
- Elaborate mechanisms and instruments for plan implementation.

Governance, organization, and linkages

- Assess the governance mechanisms of the research system.
- Review organization and structure of the agricultural research system and propose a rational and effective network of centers, institutes, and stations.
- Review and revise the mandates of the system and its components.
- Make recommendations on effective linkages with stakeholders and beneficiaries.

Program formulation

- Apply effective methods for program formulation and resource allocation.
- Determine the content of priority research programs.
- Calculate human, financial, and physical resources required for program implementation.
- Recommend procedures for program management including monitoring and evaluation.

Acquisition, development, and management of resources

- Assess the potential for obtaining research resources realistically and elaborate funding strategies.
- Establish the capital costs for refurbishing and maintaining institutes and stations.
- Determine the operational cost for implementing a research program of the size and scope proposed.

- Assess current and future availability and requirements of human resources for effective implementation of the plan.
- Propose a human resource development plan, including staffing and training policies.
- Recommend effective procedures and systems for the administration and management of human and financial resources.
- Develop an information management strategy.

The effort required to draw up a master plan of the above scope is often beyond the capacity of the few experienced planning staff in developing-country national agricultural research organizations. This is one reason why many agricultural research master plans have relied on external advisors. While this may strengthen the activity, there are risks involved in clouding ownership of the plan, increasing its cost, and perhaps undermining its relevance to local needs.

Relevance for agricultural research

Effective planning of agricultural research is crucial for contributing to national food security and sustainable agricultural development. For investment purposes, international multilateral and bilateral donor organizations require countries to prepare a specific kind of agricultural research master plan. In a number of cases ISNAR has provided technical assistance in master plan preparation as part of a donor-funded activity.

Admittedly some master plans are developed not with the intent of strengthening the capacity of national research systems by helping organizations to articulate their purpose, direction, and resources, but to respond to other political imperatives, including those of donors. Yet, rather than only "one-off" investment plans for donor-funded projects, research master plans can be effective instruments for institution building if they are designed to broaden the participation of various stakeholders and ensure that the research organization owns the processes of planning and priority setting.

Comprehensiveness is the key to both the advantages and disadvantages of master planning. Significant investments of time and financial resources are usually made in master planning, including active participation of national and international planners. The result is often a detailed outline of future activities, which may be useful for planning investments, but which may be unrealistic in a context of rapid change. Further, implementation of the plan may require more consensus building than is realized in such a comprehensive exercise. A review of master planning experiences in sub-Saharan Africa (Hambly and Setshwaelo 1996) discerned eight lessons learned:

Realism

There is a tendency to develop unrealistic systems or overly complicated targets. Furthermore, in the processes of planning and implementation, too much emphasis is often placed on quantitative information. If data requirements are beyond what the national system can sustain on its own, chances of successful completion of the planning process will be limited.

Political support

Support and commitment of national authorities and senior research leaders is crucial for successful planning, as well as for subsequent plan implementation.

Ownership of plans

The complex and technical nature of the master planning process brings with it the risk of a technocratic approach and a result that is owned neither by senior management nor by the research staff. Lack of ownership affects implementation negatively. Issues of ownership, therefore, require considerable and early attention from those responsible for the planning process. Partners such as ISNAR or donor agencies should be concerned with process management and not with substantive decision making, which is the task of national research organizations. Building consensus and creating conditions that allow the planning process to emerge and be controlled from within national institutions appear to be keystone issues.

Priority setting

In some cases, master planning documents have simply commented on what an ideal situation would be. Yet, a major task of master planning is to standardize a process for research priority setting, program identification, and monitoring that will remain in place after completion of the master planning process, through to implementation of the plan. The translation of newly set priorities into changes in resource allocations is a major challenge.

Plan implementation

Following through on implementation of stated research priorities and resource allocations, linkages, and restructuring of the organization is part of the master plan process but one which experience suggests is highly problematic in most cases. Often, more is planned than can be implemented and the detail contained in comprehensive master plans is beyond the research system's capacity to follow up. To ensure implementation, the plan must present a flexible structure and organization for national research programs. It should also comment on the degree of autonomy or decentralization of research programs, as well as the upgrading, phasing out, or development of research stations.

Linkages within and outside the agricultural research system

Master plans incorporate multiple interests of research users, taking them into consideration when setting priorities, identifying appropriate structures, allocating resources, and in implemention. The key task is to identify the role of the research organization or system under consideration vis-à-vis national and international partners, including regional research networks and associations.

Sequencing of plans

Master planning does not necessarily involve a sequential development of plans, from long-term to medium-term to annual. In practice, some countries have scaled up individual organizational plans to a broader national agricultural research plan. But this is not a general rule. Also, zonal or regional research priorities and programs should be considered as part of or follow up to master planning.

Resource requirements

Although numbers are difficult to pinpoint, experience suggests that the preparation of a comprehensive agricultural research master plan is very costly, both in terms of money and person-years invested.

Examples

Sub-Saharan Africa

Many NARS have been involved in master planning exercises. In sub-Saharan Africa, for example, nearly one-third of the countries have developed long-term agricultural research master plans (Hambly and Setshwaelo 1996). In Uganda, a national agricultural research master plan was designed for the National Agricultural Research Organisation (NARO). It includes elements such as a national agricultural research strategy; detailed research priorities along with details on how these priorities were established; descriptions of system linkages and relations with other national organizations; elaboration of the organizational structure, research networks, and support services; an overview of administration and research management; and human resource development and financial plans.

Asia

In Asia, Pakistan has received considerable international assistance in developing its agricultural research system. To improve national research planning capabilities, Pakistan prepared provincial master plans, which were then incorporated by a national steering committee into a national agricultural research master plan. The resulting plan is comprehensive in that it includes a contextual analysis of agriculture and agricultural research priorities, as well as an up-to-date research strategy. This strategy incorporates cross-commodity and commodity priority research programs, new organizational management requirements including monitoring and evaluation and information management mechanisms, a human resource development strategy, and, finally, an overview of finance requirements for implementation of the plan.

The author thanks M.M. Rahman of ISNAR for his contributions to this chapter.

References

Hambly, H. and L. Setshwaelo. 1997. Agricultural Research Plans in Sub-Saharan Africa: A Status Report. Research Report No. 11. The Hague: International Service for National Agricultural Research.

Collion, M. -H. 1991. Collaboration with national agricultural research systems in planning: a review of ISNAR experiences. Mimeo. Background paper No. 5 prepared for 1991 external program and management review, International Service for National Agricultural Research.

Ajibola Taylor, T. 1989. Tanzania National Agricultural and Livestock Research Master Plan. The Hague: International Service for National Agricultural Research.

Uganda Working Group 9A, Agricultural Policy Committee. 1991. National Agricultural Research Strategy and Plan. (Vols. I and II). The Hague: International Service for National Agricultural Research.

PARC. n.d. National Master Agricultural Research Plan (1996–2005). Islamabad: Pakistan Agricultural Research Council.

Chapter 10
Program Planning

Marie-Hélène Collion

A main task of agricultural research managers is to ensure that research programs are well defined and designed and are closely targeted to national objectives for agricultural development. These objectives constitute the starting point for program planning. Program planning is a process that requires choices and decisions to be made at different management levels for research within a particular subsector. Program planning further provides the information necessary for allocating resources among programs and the basis for formulating physical and human resource development programs for the organization. The program planning document serves as a useful tool for research managers to use in negotiating funding with government, donors, and other external partners.

What is program planning?

Program planning is a process by which the content of a research program is broadly defined for the medium to long term. Program plans further identify the resources needed to implement the program, mainly human resources (numbers of researchers and mix of disciplines), special equipment (if any), and funding for the period considered. Funding is based on a norm of operating costs per researcher, an amount that is country specific.

Program planning presupposes that research is organized in programs. A *research program* is a set of research activities, organized into *research projects* (see Capo et al., this volume) designed to address development objectives and users' needs in a particular domain or subsector. The domain or subsector covered by the program can be either a commodity (e.g., a cotton research program), a group of commodities (e.g., a food legumes program), an agroecological zone or region (e.g., Benin's central region), a production factor (e.g., farm mechanization), or a discipline (post-harvest technology or policy analysis). Note that the term "program" is also used to refer to the organizational unit that brings together researchers from various disciplines to carry out the work.

As part of a research organization's overall planning process, program planning is the step in which research priorities and programs are defined, before identifying the resources needed to implement these programs. Program planning can also be carried out without it being part of a broader planning effort. It is then a tool for systematically gathering input from program beneficiaries and indirect users, especially in times of external or internal change.

Doing program planning

Program planning is guided by an eight-step procedure (figure 1).

Step 1. Review research "domain" or subsector. The first step is a review of the state of agricultural production, marketing, and processing and of natural resource management for a particular commodity, group of commodities, region, or agroecological zone. Agricultural policy objectives, as defined by the government department or ministry responsible for agriculture, are analyzed and incorporated in the review. Situation reviews are also done for relevant production factors or disciplines. These reviews clearly identify limitations and potentialities of the domain or subsector.

Step 2. Analyze constraints. The limitations identified under step 1 are further analyzed for their causes. Constraints may relate to policy and macroeconomic factors, production factors, marketing and processing, or social and institutional factors. "Researchable" constraints, that is, constraints likely to be resolved by agricultural research, including social sciences research, are highlighted.

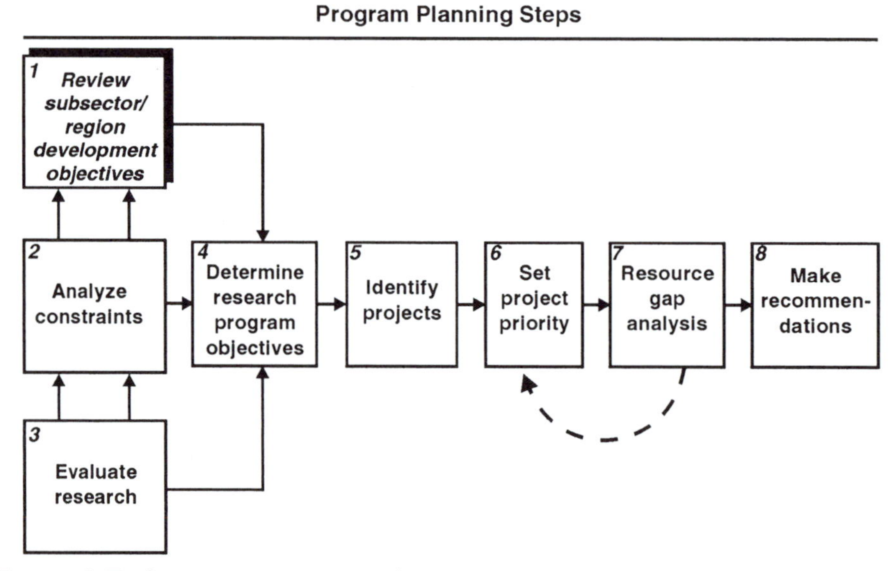

Figure 1. Eight steps in program planning

Source: Collion and Kissi 1995.

Step 3. Evaluate existing research results. Among the researchable constraints emerging from step 2, which ones have already been researched nationally, regionally, or internationally? To what degree of success? Why were they successful? Or, "What are the reasons for the lack of success?"

Step 4. Determine research objectives and strategy. Based on the analysis of constraints and the evaluation of existing research results, research objectives are set for the program. Each research objective may have a number of subobjectives, which together comprise a strategy for achieving the main objective.

Step 5. Identify research projects. A research *project* and its main activities are developed to correspond to each research objective and its subobjectives. At this stage, research projects are identified only as themes, together with an estimation of the human resources (disciplinary skills and researcher time) needed to do the work.

Step 6. Set priorities among research projects. Typically, more projects are identified in step 5 than can be implemented. The ones most important for achieving overall program aims are selected in this step. To do so, criteria are needed to measure each project's potential contributions to achieving the overall program objectives. Also, a methodology for combining the projects' ratings for each criterion is needed (see Contant, this volume). Priority setting results in establishment of a group of essential projects.

Step 7. Human resource gap analysis. The aggregate human resources required to implement the group of priority projects are identified in this step. By comparing the human resources needed with those available, it becomes clear where programs need strengthening. If a human resource gap cannot be filled, because the resources for hiring or training staff are unavailable, then the group of priority projects has to be redefined.

Step 8. Recommendations for implementation. In this final step, measures are determined for making the program operational and ensuring that program results will have the intended impact on development. These recommendations stem from non-researchable constraints that emerged in the analysis, which can have an important impact, for example, on the level of adoption of research results.

Participation

Program planning is an activity that brings researchers together with the beneficiaries of research results (first and foremost producers, but also processors, traders, and exporters) and indirect users, such as extension workers and policymakers. Beneficiaries and indirect users should provide input in the subsector review and be closely involved in constraints analysis, evaluation of existing research results, identification of research objectives, and setting of priorities. In addition to ensuring program relevance, stakeholders' participation in program elaboration fosters commitment to program implementation: after having been involved in planning a research program that responds to their concerns and needs, stakeholders will be on the look out for results.

The best way to bring together these different perspectives is through workshops at which representatives of all beneficiary groups are present. Careful selection of participants is essential to assure quality and relevance of program elaboration. Effective farmer representation is difficult to achieve. Farmers appointed by producers' organizations generally represent farmers' perspectives better than individual farmers do, provided that the organizations and their leaders can speak for the diversity of production systems and socioeconomic circumstances relevant to the program being planned.

Organization

Program planning should be the responsibility of an ad hoc planning group under the program leader, assisted by one or two senior researchers from the program and a staff member from the organization's planning unit or from the scientific directorate. The presence of a professional planner ensures that the planning methodology is used consistently among programs within an organization.

As indicated above, the program planning exercise itself can be organized around two workshops, with a final review meeting. The first workshop, a three-day event, covers steps 1 to 4 and produces a set of research objectives. Between the first and the second workshop the planning group identifies research projects. Background data needed to set priorities is collected and analyzed. During the second workshop, research projects are prioritized, leading to the identification of a set of projects considered essential for implementation. Recommendations for implementation are also formulated. After the second workshop, the planning group writes the program document, which is submitted for external validation during a review meeting.

Inputs for program planning

Information should be collected on commodities, production systems/factors, natural resources, and national or regional development objectives. Useful data on commodities includes area under cultivation or size of herds; quantity produced, marketed, and processed and trends; contributions to agricultural gross domestic product (GDP) and rural employment; estimates of future demand due to population or income growth; supply and demand elasticity; vulnerable or target groups (for food security or equity considerations); trends in imports and exports; and potential for income and foreign exchange earnings. This data should be disaggregated by agroecological zone as far as possible. Also useful may be basic physical data (climate and soils), socioeconomic characteristics of producers (number and size of farms, management levels, and input use), and productivity (actual and potential, on station as well as on farm).

Research results produced within the country and beyond its borders should be well documented: what is the level of adoption of the research results already transferred? If results were not well adopted, what are the reasons? Why is some research considered unsuccessful? Some of this information will be available from various reports. Workshop participants' knowledge and experience is an important source of additional information. Choosing participants who are knowledgeable about the subsector is therefore crucial.

Tools for program planning

A number of planning tools can support the planning exercise (see Part IV of this volume). Two examples are the logical framework and the ZOPP method, the latter of which involves construction of hierarchies of constraints and objectives. Techniques for managing group discussion and reaching consensus, such as visualization or Delphi techniques, are also extremely useful. Finally, the planning group must choose from a number of priority-setting tools which one is most appropriate for setting priorities among research themes in their specific context. In most cases, a simple version of a cost-benefit analysis may be the most reliable and fairly easy to handle (Collion and Kissi 1995).

Relevance for agricultural research

Program planning is an essential tool for ensuring research relevance and consistency within a program. It brings together researchers, beneficiaries, and indirect users for a joint definition of program content, taking into account producers' needs as well as national or regional development objectives. It promotes communication among researchers, producers, development workers, and policymakers and enlists commitment, especially that of policymakers, for implementation and eventual resource allocation. It is furthermore an excellent team-building exercise for scientists in the program.

Within the research organization, the document that results from the program planning exercise serves primarily as a tool for program management: projects proposed by researchers must correspond to the priority research "themes." The document provides guidelines on allocating resources among projects and the basis for formulating physical and human resource development programs.

Lastly, the program plan is a valuable tool for managers to use in negotiating funding with government, donors, and other external partners. By specifying the results that can be expected, the document strengthens their case in requesting funding.

Examples

Most research organizations have been involved in program planning. Examples in Africa include Morocco's Institut National de Recherche Agronomique (INRA), Benin's Institut National des Recherches Agricoles du Bénin (INRAB), and the Kenya Agricultural Research Institute (KARI). In Asia, the Centre for Agricultural Research Programming of Indonesia supports some 16 research institutes in elaborating their research programs once every five years; and in South America, Brazil and Argentina have conducted program planning.

Morocco

In Morocco, a cycle of program planning was initiated in 1990 and has focused mainly on commodity program planning. One example is the legume program with its faba-bean subprogram. The resulting program plan describes the faba-bean subsector in Morocco and the groups targeted by the research program, small-scale subsistence farmers and poor urban consumers. It goes on to analyze future demand based on population growth and income elasticity of demand and the country's development objectives (target area under faba-bean cultivation and target farmers' yields). Constraints analysis for the program planning was done by constructing a hierarchy of constraints. Researchable constraints were then identified and research results and their adoption rates analyzed. The next step was the identification of research objectives (using the ZOPP method). Twenty-eight research projects were identified. Priority setting, using a cost-benefit analysis, helped workshop participants determine which projects were essential for achieving program aims. A human-resource gap analysis was carried out, identifying the need to redeploy or recruit an economist, a weed scientist, an entomologist, and a post-harvest specialist.

Recommendations were also made for implementation. Two workshops were held, with participants from universities' faculties of agriculture, staff from directorates of the Ministry of Agriculture and Agrarian Reform, staff from the Agricultural Credit Union, representatives of the Federation of Cereal and Leguminous Crops Producers, and scientists from European universities.

References

Collion, M. -H. and A. Kissi. 1995. Guide to Program Planning and Priority Setting. Research Management Guidelines No. 2. The Hague: International Service for National Agricultural Research.

Dagg, M. and F. Haworth. 1988. Program Formulation in National Agricultural Research. Working Paper No. 17. The Hague: International Service for National Agricultural Research.

Janssen, W. and A. Kissi. 1997. Planning and Priority Setting for Regional Research. Research Management Guidelines No. 4. The Hague: International Service for National Agricultural Research.

Kamau, M. W., D. W. Kilambya, and B. F. Mills. 1997. Commodity Program Priority Setting: The Experience of the Kenya Agricultural Research Institute. Briefing Paper No. 34. The Hague: International Service for National Agricultural Research.

Macagno, L., J. Pizarro, and G. Cordone. 1992. Caracterización de los programas de investigación: un aporte metodológico para la fijación de prioridades. Documento de Trabajo. Buenos Aires, Argentina: Instituto Nacional de Investigación Agropecuaria: Dirección Nacional de Planificación.

Chapter 11
Research Project Planning

Olga Capo, Silvia Galvez, and Ron Mackay

Project planning in agricultural research is a systematic and integrated management approach to identifying and preparing a plan to resolve a "problem" identified within the broad field of agriculture. Like the other types of planning mentioned thus far in this booke, project planning involves beneficiaries and other research stakeholders (end users). They are essential players in determining the optimal way to solve a problem in the sector using research. Ideally, projects should contribute to higher level program and organizational objectives. However, they do not always do so. A few large externally funded projects may actually divert resources away from established priorities.

What is research project planning?

Agricultural research project planning is a systematic and integrated management approach to identifying and preparing a plan to resolve a "problem" identified within the broad field of agriculture. The problem may relate to averting a crisis, meeting a need, or grasping an opportunity in the production, management, harvesting, storage, processing, or marketing of a commodity, crop, animal, or other natural resource. Effective research project planning is a critical initial phase of the larger cycle of research project management. It encompasses phases 1 through 4 of the management cycle in table 1.

Table 1. Research Project Planning Within the Research Project Management Cycle

Phase 1	Phase 2	Phase 3	Phase 4	Phase 5	Phase 6
Project area and objectives identified	Research proposal prepared	Research proposal revised	Project objectives and finance approved	Project execution and monitoring	Project evaluation

Source: Gapasin 1993.

A project differs from continuous routine operations in the sense that it is a temporary undertaking to create a unique product or service. Projects have specific begin and end dates and they produce specific outputs or results.

Like the other types of planning mentioned thus far in this book, project planning involves beneficiaries and other research stakeholders (end users). They are essential players in determining the optimal solution to a research need and the best means to arrive at it. A multidisciplinary research team is frequently required to achieve targeted project outcomes. Researchers and end users are engaged in each step of project planning, from the specification of research objectives to the elaboration of methodology, including procedures and resources that will lead intended users to adopt project outputs, and to setting criteria for determining project success.

Projects should contribute to higher level program and organizational objectives. To achieve this, a coherent set or "portfolio" of projects is designed. Planning tools such as logical frameworks help ensure a structured approach to program planning and to make sure that different projects are individually necessary and collectively sufficient to achieve program objectives. Projects do not always contribute logically to programs, however. A few large externally funded projects may actually divert resources away from established priorities.

Doing research project planning

The research project planning cycle is a critical one because it launches the project "start-up" phase. The more thoroughly and conscientiously it is carried out, the greater the likelihood of the project's ultimate success. Research project planning has three dimensions: process, content, and context.

Process

There are no set rules for the different steps in research project planning. Processes range from very simple to elaborate, depending on the size of the project and requirements of the organization or the research financier. Often the following steps are relevant, although not necessarily in this order.

Step 1. Preparation of a project idea or profile. A project idea arises from experienced or potential needs at the local, national, regional, or international level. These needs are the product of current and ongoing challenges facing the agricultural sector. Relevance of the project idea is first confirmed with the constituency experiencing the need. The project idea must make an explicit statement of the problem to which the project will be directed and provide a precise summary of who the primary users of the results will be. The project idea document should not exceed one page in length.

Step 2. Initial technical and management review. Using feedback obtained from peers and supervisors during step 1, the project idea is reviewed. The objective of the review is to submit a stronger and more elaborate project document. The review should assess the relevance of the project idea; its scientific, technical, and methodological aspects; and funding implications. Particularly for large projects it is important that this review be conducted at a relatively early stage to confirm that the project idea falls within the parameters of the strategic plan of the organization. As preparation of full project documents is a costly affair, it is important to either eliminate or redirect project ideas that do not (fully) meet technical and management requirements. At this stage, the project manager should become aware of the requirements of different donors regarding content and format of project proposal documents.

Step 3. Preparation of full project documents. When the merit of the project and its congruence with the organization's strategic plan has been confirmed, the researcher who will become the project manager drafts the definitive project proposal. This step involves an analysis by the project manager of the background conditions that exist with respect to the research topic, that is, the state of the art as revealed by a thorough literature review. A complete and thorough review of the current state of knowledge on the topic is essential for confirming the project's viability. The need for the project must be clearly justified, its products defined, and the advantages for the identified beneficiaries spelled out. A description of exactly how project outcomes will resolve the problematic situation identified at the outset also needs to be provided.

Tasks to be undertaken in order to complete the project and the sequence in which they must occur are also specified. In large projects this may constitute a list of subprojects or experiments, perhaps presented in a "work breakdown structure." The anticipated time required for each activity is hereby calculated and justified. How the successful completion of each activity contributes to the successful attainment of the project's objectives and goal is also clearly spelled out.

Confirming availability of necessary resources (staff, funds, facilities, and equipment) is an important part of preparing the full project document. Resource requirements and especially finance usually need to be determined precisely, justified, and presented in an appropriate format. Funding for the approved project may be sought internally within the national agricultural research system or externally, from competitive national, regional, or international funding sources. Research project planning and the various formal stages through which the research proposal must pass should be carefully managed in conjunction with externally imposed deadlines for proposals so that projects may be submitted to the appropriate funding body on time.

Step 4. Project review and approval. When the completed project documents have been accepted internally within the organization or unit where they originated, the project might need to be presented at a higher level for formal approval. This second-tier of approval may have both technical and financial dimensions. Approved projects are those deemed to be technically and procedurally adequate and conform to strategic plans at all levels – departmental, research center, and research institution. If the project was designed for external funding, it may now be submitted to a prospective donor. Once the budget is procured from a funding source, financing arrangements must be duly approved and a contract with the funding source formalized by the organization. The budget of the project, supplemented by a schedule or time frame for activities, completes the initial project planning phase (table 2).

Content

Ideally, agricultural research projects share the following features:

- a clearly identified problem facing specific beneficiaries and stakeholders
- a solution acceptable to the end users
- a precise project objective
- an explicit research methodology
- a clear set of facilitating activities
- one or more outputs that are visible and specific
- a clearly specified use to which the outputs will be applied
- identified project milestones and a monitoring and evaluation process
- a time frame or schedule of activities
- a budget

Problem definition is backed by a thorough review of literature and existing research results. It may also be addressed using a SWOT (**s**trengths, **w**eaknesses, **o**pportunities, and **t**hreats) analysis or constraints analysis.

Context

Agricultural research projects are planned within a very specific institutional context. This context may include factors such as national and international economic changes, new policy regimes, declining amounts of research funding

Table 2. Milestones in research project planning

Milestone 1	Milestone 2	Milestone 3	Milestone 4	Milestone 5
Identification and confirmation of project idea	Delivery and acceptance of project profile	Delivery and acceptance of project proposal	Obtainment of project approval	Procurement of project funding

from government, the obligation to bid competitively for research funds, and an increasing role of the private sector. Structural changes within the research organization itself may also have been made to match changes in its environment, national role, goals, strategies, technologies, and the evolving skills of its personnel.

Research project planning is one mechanism for coordinating research efforts while structural and other changes are being experienced within and around the organization. Concrete steps towards standardizing the research planning process helps to ensure that as research activities become more diversified and complex, a necessary level of coordination and control is maintained.

A collaboratively designed standardized process for research project planning within any organization ensures that proposals are developed in a form optimal for that organization. Appropriate, standardized proposals and accompanying processes for reviewing them at various critical stages ensure that projects closely match the organization's goals, environment, technology, professional expertise, and strategy. Rational project planning also encourages establishment of linkages with relevant beneficiaries and stakeholders to ensure maximum research relevance and cost-effectiveness.

Together, the process, content, and context of agricultural research project planning constitute the initial phase of the research project management cycle. Table 3 illustrates the key documents and collaboration required for research project planning.

Relevance for agricultural research

Sound research project planning is key to transforming a "problem" situation into a "satisfactory" one. The process of project planning is designed to provide the best solution to a significant problem facing a national agricultural research organization and its clients and to facilitate the adoption and effective use, monitoring, and evaluation of that solution.

The process of research project planning described here is a strategic, participatory approach, sensitive to the organizational context, as well as the social and economic environments, to satisfy the broader range of local, national, and regional stakeholders, including producers and government. Effective agricultural research project planning also ensures a strong element of competitiveness in the organization to maximize the likelihood of capturing funds and challenge to researchers.

Examples

Chile's national agricultural research institute "INIA" is responding to changes in its external environment. By means of a rational, comprehensive, and incre-

Table 3. Research Project Planning: Key Documents and Level of Collaboration and Approval Required

Project Idea Document	**Project Profile Document**	**Project Proposal Document**
Level of collaboration / approval: 1. intended user 2. research peers 3. head of department	*Level of collaboration / approval*: 1. intended user 2. head of department 3. director of research	*Level of collaboration / approval*: 1. intended user 2. institutional research committee
Title	**Title**	**Title**
Type of project	**Type of project**	**Type of project**
Area of project	**Area of project**	**Area of project**
Scope of project	**Scope of project**	**Scope of project**
	Location – of project operations – of project impacts	**Location** – of project operations – of project impacts
		Summary
		Current state of research
Objectives – general – specific	**Objectives** – general – specific	**Objectives** – general – specific
Rationale	**Rationale** – problem situation – resolved situation	**Rationale** – problem situation – resolved situation
		Methodology – transfer mechanisms – critical elements of the research methodology
	Potential research team	**Potential research team** – participating departments – disciplines required
	Disciplines required	**Products anticipated (associate each with a specific objective)** – scientific/technical – economic – socioeconomic
Duration of project	**Duration of project**	**Duration of project** **Provision for monitoring and evaluation** – context – inputs (physical, human, financial) – processes – products – impacts (economic, social, scientific)
	Resources required – material – human	**Resources required** – material – human
Total budget	**Total budget**	**Total budget**
	Potential sources of funding	Funding source for submission

mental approach to research project planning, INIA has addressed challenges including changes in national and international economies, new strategies employed in international competitive marketing of agricultural commodities, changes in national patterns of consumption and demand, growth of agro-industry, increasing appreciation of the heterogeneity of Chile's agroecological reality, the need to replace imports with local production, and the critical dollar-value earning capacity of Chile's agricultural exports.

At the same time, by effecting internal changes in operating procedures, INIA has brought its mission in line with national economic and food-security requirements. Redesigning its organizational structure to fit its changing circumstances, the institute has decentralized substantial budgetary authority to its 12 experiment stations and eight regional research centers, enabling them to respond effectively and efficiently to the wide range of location-specific requirements in the country. INIA's revenues have evolved from almost total dependency on the agricultural ministry to more widely diversified sources of funding. These include national competitive research funds, regional development funds, and contract funding from agroindustry.

These dramatic changes have challenged INIA to ensure regional relevance while preserving a broader, national vision. New coordinating mechanisms were needed to guarantee unity of purpose. The core strategy adopted to enable management and research personnel to cope effectively with increasing uncertainty, as well as decentralization, has been to increase capacity to generate, share, and process relevant information at all levels of the organization.

Further, the "research and development project" was identified as the unit of management. A project-planning mechanism was created to promote standardization of information in the absence of direct central supervision and in acknowledgment of the impracticality of informal communication in such a large and agroecologically diverse country.

Regional centers use the standardized research planning process to develop projects in response to local needs within the goals of their regional strategic plan. The regional plans were developed to ensure congruity with INIA's overall strategic plan. A national review committee on which central management and all regional research directors are represented reviews all project profiles (the earliest stage of the project plan). This review task is simplified by a standard format agreed on by management and research personnel. The national review committee identifies the strengths and weaknesses of each proposal, as well as real or potential areas of overlap. It suggests where collaboration between centers might promote economies of scale and eliminate potential duplication of effort. It further identifies resources, such as research skills, that can be shared among projects. Finally, the review committee locates appropriate funding bodies to which specific projects can be addressed to maximize prospects for funding. After project profiles are modified, if necessary, and approved, a full research proposal is prepared, revised, and submitted for final approval. Once funding has been secured and a budget allocated, the project starts and is monitored regularly to ensure its compliance with the plan. Monitoring may also show that revisions in the plan are needed. Each project is evaluated upon completion.

INIA's planning steps and its formal, standardized project planning format provide information agreed on as essential by management and researchers alike. The information is aggregated at different levels of management (i.e., at the levels of the project, program, regional center, and central management at headquarters). Availability of appropriate information at each level helps ensure adequate and effective coordination, resulting in improved efficiency of resource use and relevance of research projects.

References

Aued, J. et al. Competitividad en proyectos: el triángulo de la competividad en proyectos. Proyecto "Fortalecimiento de la Planificación, Seguimiento y Evaluación en la Administración de la Investigación Agropecuaria en América Latina y el Caribe". The Hague: International Service for National Agricultural Research.

Bulo Z., I. Fremy, O. I. Capó, S. A. Gálvez and Phillips F. Ronnie. Sistema de aprobación de proyectos de investigación y desarrollo. Agosto 1997. Chile: Instituto de Investigaciones Agropecuarias, Dirección de Estudios Planificación y Proyectos.

Collion, M. -H. and A. Kissi. 1995. Guide to Program Planning and Priority Setting. The Hague: International Service for National Agricultural Research.

Gapasin, D. 1993. Research project management. In *Monitoring and Evaluating Agricultural Research: A Sourcebook* edited by D. Horton et al. The Hague: International Service for National Agricultural Research.

IARC/NARS training group for sub-Saharan Africa. 1999. Planning, Monitoring and Evaluation for Research Projects. Mimeo. The Hague: International Service for National Agricultural Research.

Locke, L. F., W. W. Spirduso, and S. J. Silverman. 1993. Proposals that Work: A Guide for Planning Dissertation and Grant Proposals. Third edition. Newbury Park: Sage.

Chapter 12
Experiment Planning

Jörg Edsen

Experiment planning represents the lowest level of agricultural research planning. In comparison with higher levels of planning, it places more emphasis on achieving technical quality in terms of experiment design. Experiment planning aims to identify the most efficient and effective option for achieving the research results necessary for developing a required technology. An important outcome of the experiment planning cycle is a technically, scientifically, and statistically sound experiment proposal that details of a prospective course of research. Influential factors such as natural cycles, financial rules, and higher level program or project implementation schedules must be considered. A well defined, standardized process of experiment planning helps ensure the fair and equal treatment of all research ideas within the organization and eases communication and coordination of research planning between the organization's headquarters and research stations.

What is experiment planning?

Agricultural research is generally organized in a hierarchical structure of decision-making levels. Directly concerned with conducting research are the levels of programs, projects, and experiments. Higher levels of decision making include the levels of organizations, units, or stations (Horton et al. 1993). At the level of experiments, the actual research work is done, and field trials, studies, surveys, and other types of investigations are carried out.

During the experiment planning process, the results of project planning (project objectives, a broad outline of project activities, priorities) are taken up and used to develop the details of experiments. These details are described in an experiment proposal that constitutes the first document of a prospective experiment. The experiment proposal reviews experiments done elsewhere on the same or similar topics, as well as the results obtained elsewhere. Experiment proposals usually contain a summary description of the experiment, its objectives, rationale and justification, research methodology, budget, specific activities, resources, and a detailed work plan.

Sound and effective experiments are crucial for any research organizations that aims to produce new knowledge and high-quality technologies. Whereas political and policy concerns are important for long-term strategic planning, for experiment planning technical insights play the prominent role. Experiment planning is strongly grounded in the theories of "experimental design," part of the discipline of biometrics.

Researchers and managers at different levels are the main players in the experiment planning process. Herein, intimate knowledge of the research subject and of experiment design are combined with an understanding of management of the experiment planning cycle. Researchers and managers interact in a number of ways to design experiments and review experiment proposals. Review of proposals leads on to a decision on whether the experiment is to be undertaken. Duration of the process, from the first planning initiatives to fund allocation, varies considerably, depending on the institutional environment and funding agents' rules, regulations, and deadlines.

Doing experiment planning

Managing experiments follows a process similar to the research management approach described by Gapasin (1993). She identifies six steps that, in principle, can also be applied in managing experiments. Only the first four steps relate to planning:

1. identify experiment areas and objectives
2. prepare experiment proposals
3. review experiment proposals
4. approve experiments and resource commitments
5. implement and monitor experiments
6. evaluate completed experiments and their impact

The following sections examine the first four of these steps, those related to planning, defining them more precisely for practical application.

The experiment planning cycle

Figure 1 shows a cyclical sequence of nine events in planning and approving experiments. In this example, the financial year starts in July, with submission of the budget to government in end December. Note, however, that the planning cycles of individual research organizations may vary greatly, depending on their individual environment. Some general factors that influence the planning cycle are discussed below.

Institutionalizing an experiment planning cycle presupposes that planning has taken place at the project level. The objectives, broad activities in the parent project, and priorities among the activities need to be known. If these details are

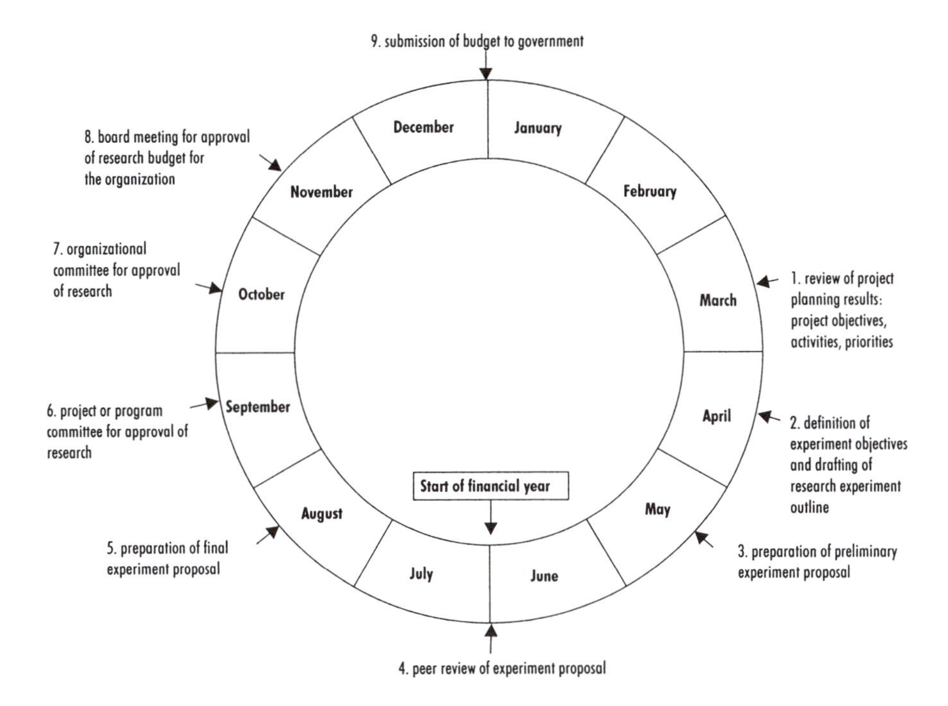

Figure 1. An experiment planning cycle for agricultural research

unavailable, the approach to planning at the project level may first need to be revisited and adapted. The principles of project planning are described in Capo et al. in this volume.

The first event of the experiment planning cycle as shown in figure 1 is a review of the priorities and objectives of the parent project. This results in the definition of specific objectives of the experiment and the drafting of an outline for the experiment (step 2). The outline is drawn up by a group of project scientists or subject-matter specialists.

A comprehensive experiment proposal is then prepared by a designated scientist in consultation with colleagues. A review of experiments done elsewhere is central to this stage of proposal drafting, to avoid duplication. Preparing an efficient and effective experiment design is the other key activity in this step (3). Next, a technical committee of peers or external specialists reviews the proposal to ensure that it is scientifically and statistically sound and technically feasible (step 4). Comments and improvements are incorporated into the experiment proposal by the designated scientist (step 5).

The resulting final version of the experiment proposal is discussed by a project or program committee (step 6). This committee reviews all experiments proposed for the project or program. It decides on resource allocation and on a schedule of research activities for the coming financial year. Budgets and individual work plans for experiments are merged into one document describing the project or program activities and resource requirements for the next financial year. The document may be discussed in higher level approval committees, for example, an organizational approval committee (step 7) or a senior management approval committee (step 8). Finally the budget of approved research experiments is included in the organization's total annual budget. This budget will eventually be submitted to the government and for inclusion in the national budget for the coming financial year (step 9).

Factors influencing the experiment planning cycle

As the experiment planning cycle described here is not universally applicable, the experiment planning *process*, specifically the timing and type of events, must be adapted to the individual situation of the research organization. In doing so, factors that influence or are closely linked to the research planning process need to be taken into account. A few examples can be mentioned that are particularly relevant to agricultural research:

1. natural cycles, like the cropping cycle or other seasonal cycles
2. financial rules and regulations of the agencies that fund research, for example, begin and end of the financial year, deadlines for budget submission, disbursement schedules
3. implementation and conduct of research planning at higher levels, for example, at the program and project levels

Some or all of these factors will influence the research planning process. Researchers and managers have to be aware of these factors and adjust the planning processes in their organizations accordingly.

Natural cycles. Agricultural research deals with natural processes, which are either the subject of the research or which exert great influence on the actual implementation of the research. Such natural processes may repeat themselves annually or in specific natural cycles. The best known cycle is the annual cropping cycle, which depends on rainfall patterns. Other cycles include annual migration patterns of livestock, wildlife, and fish and recurrent pest and disease calamities for livestock and crops. In many cases, natural phenomena determine the timing of research implementation. For example, most cropping experiments must commence at the start of the rainy season. Other agricultural research activities, including laboratory experiments and social studies, may depend to a lesser extent on these natural phenomena.

Consequently, the best time for experiment planning may be four to six months before implementation of the experiment is expected to begin. Unfortunately, the regime of natural cycles often cannot be fully considered in experiment planning, due to other factors that influence implementation to an even greater extent, such as financial rules and regulations.

Financial rules and regulations. In the day-to-day management of a research organization, it is unfeasible to set up an individual planning schedule for each research activity. Detailed planning of research is therefore often done in a combined effort by researchers and managers at one specific time of the year. This "research planning season" depends on the financial year and the financial rules and regulations set by the agencies that fund research (ministry of agriculture, other ministries, private organizations, donors). Financial rules regulate when budgets must be submitted and when disbursements are made. Subsequently, the research planning process has to be adjusted in such a way that the specific deadlines can be met. This may mean that some research activities need to be planned well ahead of their actual implementation.

Higher level planning. Although it may not always be done in practice, effective program and project planning is a prerequisite for planning experiments that fit into a coherent research portfolio (see also Collion and Capo et al., this volume). The work plan, the outputs, and the objectives of the experiments have to fit logically into the project and program structure of the organization. Also, the budget for the experiment portfolio has to match the resources available for the respective projects and programs.

To allow for a logical build up of the research agenda, the timing of planning events for program, project, and experiments needs to be coordinated. The best scenario, which cannot always be adhered to, would be to conduct program planning first, then project planning, and finally experiment planning. However, only experiment planning is conducted annually. As a rule, project planning occurs every three to five years and program planning every 10–15 years. Adjustments may be made at all levels when necessary, provided they are coordinated with the respective higher or lower levels.

The factors discussed above are related; each may influence the experiment planning cycles of a particular research organizations to varying degrees. The factors' potential impact, therefore, has to be assessed in the environment of the research organization and in relation to the other factors. Researchers and research managers need to be highly flexible in designing and applying the experiment planning cycle.

Relevance for agricultural research

The main outcomes of the experiment planning process are technically and scientifically sound experiment proposals. The subject of the experiments should be relevant to the beneficiaries of research, and the results should contribute to the achievement of the objectives of the parent project. Implementing a sequence of related planning events within the research organization helps strengthen the planning process and fosters a more coherent overall research agenda. The events in the process provide a forum for decision making and prepare the ground for successful implementation of field trials, studies, surveys, and other investigations. Experiment planning should ensure that experiment proposals have realistic budgets and annual work plans and that the available resources are efficiently allocated.

Some research organizations have no explicit procedure for planning experiments. Guidance for scientists and research managers and supervision of the process by the planning unit is often difficult due to understaffing, problems in communication, and limited funds. This often results in a variety of planning approaches being used in programs and research stations within the same organization.

Detailed instructions and guidance on how and when to conduct the planning events help ensure quality, consistency, and timeliness of planning throughout the research organization. Developing an experiment planning cycle is a first step towards standardizing experiment planning throughout the research organization. A clearly defined, standardized process ensures the fair and equal treatment of all ideas and allows improved coordination of research activities between headquarters and research stations. Beyond establishing a logical sequence of planning events and deciding on their timing, other measures can improve experiment planning. Developing standard proposal formats, establishing common criteria to measure and evaluate proposed research, and giving clear instructions as to how different planning events should be conducted constitute valuable accompanying measures. They make experiment planning easier for scientists and more effective for research managers. Moreover, a well designed experiment planning process fosters confidence among donors in management of the research organization and in their efficient and effective use of funds. This may result in increased funding and deter funding agencies from imposing their own planning procedures.

Examples

Uganda

Uganda's National Agricultural Research Organisation (NARO) undertook the streamlining of experiment planning at its research stations. Several factors prompted the effort. First, the World Bank, as a major NARO donor, requested that it to improve research planning. Second, NARO had implemented a management information system (MIS) that needed to be integrated into planning, monitoring, and evaluation (PM&E) and thus necessitated a more rational planning approach. Last, a research priority-setting exercise that had been undertaken required follow-up planning actions, in order to harmonize NARO's research agenda with its newly established priorities.

NARO applied a learning-by-doing strategy to the challenge of enhancing planning. Especially challenging was that NARO needed to review and plan its research agenda at all levels: programs, projects, and experiments. During design and implementation of the planning process, making a clear distinction between the different levels and their associated planning tasks proved beneficial.

Kenya

Learning from NARO's experiences, the Kenya Agricultural Research Institute (KARI) began to make improvements in their planning by focusing only on the experiment level of the research hierarchy. Another dimension of difficulty was added when KARI decided to streamline its monitoring and evaluation of experiments at the same time.

KARI had already made several attempts to institutionalize a standard system for monitoring and evaluation throughout the institute. While the procedures were known in theory, apart from some common features and events, research monitoring and evaluation was conducted in a variety of ways in KARI's different programs and centers. This prompted donors to impose their own PM&E procedures. Furthermore, although KARI had implemented an MIS, the variety of PM&E approaches used made the full implementation of the system impossible. The unit responsible for the MIS eventually took the lead in standardizing PM&E for experiments.

In both Uganda and Kenya, the institutes' planning units played key roles in the process (in NARO the Monitoring, Evaluation and Planning Unit and in KARI the Socio-economics Unit). At the time of this writing, NARO continues to consolidate its planning and has started streamlining monitoring and evaluation. KARI has initiated an organization-wide process for streamlining PM&E.

References

Gapasin, D. 1993. Research project management. In *Monitoring and Evaluating Agricultural Research: A Sourcebook* edited by D. Horton et al. Wallingford, UK: CAB International.

Horton, D. et al. (eds). 1993. Monitoring and Evaluating Agricultural Research: A Sourcebook. Wallingford, UK: CAB International.

Chapter 13
Planning Financing and Investment

Gary Alex and Derek Byerlee

Financial planning for research organizations aims to reconcile the level of research activity with the likely availability of funds. Financial planning requires strategies for identifying and developing alternative sources of funding; using and allocating funds in the most efficient manner possible; and adjusting program and organization size to the projected funding base. Financial planning requires short-term planning through the annual budget cycle, as well as long-term strategic financial planning, although the latter often receives less attention.

What is financial planning?

The approach to financial planning for agricultural research differs according to the level at which it is conducted. At the broadest level – that of the national agricultural research system – financial planning involves formulating national policy for agricultural research funding. This requires responses to questions such as "What level of public funding should be allocated to research?" "What mechanisms should be used to allocate these funds?" And, "What are the priorities?" Further, means should be established for encouraging private funding of research. At the level of a given research organization, whether it be a national research organization, a university, or a commodity research organization, financial planning is similar to that done in a commercial firm, requiring that projected income be balanced with projected financial needs for a specific planning cycle. This chapter focuses on financial planning at the research organization level, emphasizing the medium term of three to five years.

Financial planning has two principal dimensions: planning expenses and planning how to develop and diversify funding. In the past, when most funding for (public) research came from one source (a ministry or treasury), attention to the former was often adequate, and financial planning with emphasis on personnel and operational costs and investments sufficed. Now, however, the latter dimension of planning is growing in importance. This requires assessments of the types of funding sources that could be developed, as well as changes in the way that expenses are planned.

An organization's financial plan is summarized in a budget which details the resources allotted to each research program and to maintaining future research capacities. Typically extensive analysis and numerous supporting documents are required to substantiate the budget tables. Preparing financial plans is becoming the major task of research managers, requiring allocation of considerable management time for raising funds, defending budgets, allocating resources, and planning expenditures.

Sound financial planning is critical to a research organization's success. The financial plan is the basis on which strategies and programs are implemented. Without a sound financial plan, an organization is unlikely to be able to obtain financing for its programs or use resources efficiently to accomplish strategic objectives. The financial plan is also an essential part of an organization's strategic plan. Without a realistic projection of funding availability and allocation, a strategic plan may be only a wish list. Financial planning must consider both short-term, annual budgets and long-term budget projections.

Short-term financial planning is accomplished through an efficient budget cycle. This cycle provides guidance to program managers in their preparation of budgets that allow for efficient implementation of annual work plans. Since substantial funding usually comes from government sources, the organization's budget cycle is typically linked with the national budget process. Preparation of annual budgets requires projections to be made of annual operating costs and resources to be allocated between operating costs and long-term capacity building and maintenance. Decisions on resource allocation call for a long-term perspective and understanding of strategic plans and financial prospects. Unfortunately, short-term financial pressures often prevent research administrators from focusing on the strategic financial planning needed for the long-term development and productivity of an organization.

Most public research organizations are dependent on annual government budget allocations and may not even have budget projections for more than a year or two. Acquiescing to such a system is short sighted, however, and likely to limit the organization's future potential. Whether formal or informal, a research organization should have a five- to 10-year projection of funding requirements and sources of finance. Such a perspective enables managers to be more proactive in developing alternative sources of funding, ensuring development and maintenance of human and physical resources, and reconciling program size to expected resource availability.

Doing financial planning

At the strategic level, a research organization's long-range financial plan must be developed within the context of an overall strategic plan. This is, by necessity, an iterative process. A draft of the organization's vision and strategic plan should come first and then be subjected to the rigor of a realistic assessment of likely availability of funding from different sources. Generally, this will leave a financing gap, which necessitates revisions to the strategic plan, additional efforts to raise funds, or both.

Sources and the level of financing available to a research organization influence both the research agenda and the scale of the strategic plan. If, for example, a substantial share of the budget comes from environmental groups or a commodity group, the strategic plan and research agenda will have to be seen as responsive to the needs of that group. On the other hand, a research organization should not allow an individual source of funds to hijack the organization's agenda so that, for example, research on commercial crops crowds out research on subsistence food crops.

In practice, financial planning must start from the organization's current budget. Unless there are special circumstances such as the start or completion of a large project, rapid changes in a budget are unlikely, especially when funds come from government sources. When current funding and resource needs differ greatly (as they often do!), management is faced with an important fund-raising agenda.

The annual budget cycle draws longer term financial plans into the reality of the organization's annual budget (see Bruneau, this volume). The budget cycle may need to begin as early as 18 months prior to the start of a fiscal year with a request from the budget office to research program leaders and other operating units for projections of funding needs in the following year. Research program leaders should be provided with "high," "expected," and "low" budget envelopes and asked to prioritize activities within these scenarios. Projections from operating units are consolidated, rationalized, and returned to the programs for further comment or justification. Several iterations are usually required to finalize a budget for presentation to the organization's board or other funding authorities. Good financial management requires the budget cycle be linked to organizational monitoring and evaluation practices. This enables funds to be allocated based on performance and productive programs rewarded.

Developing and diversifying funding sources

Identifying and evaluating potential sources of funding is the first step in developing and diversifying funding sources. Although government budgets are – and will likely remain – the principal source of funding for most agricultural research, public research organizations are well advised to make efforts to diversify sources of funding, in order to increase both the level and stability of funding and to intensify relations between research programs and clients.

Increasingly, funding can be divided into "core" funding and "project" funds, with the latter resources earmarked for specific activities. Core funding is vital for an organization to maintain its strategic focus. Project funding, although it may contribute to implementation of the organization's strategy, is less flexible and often less closely targeted to the organization's explicit goals and objectives. If project funding exceeds about 35% of an organization's operating budget, the organization may be unable to maintain its independence and strategic focus.

Some major funding sources for consideration are government block grants, competitive contracts or grants, joint ventures and other private-sector collaboration, commercialization of products or services, farmer funding, endowments, and donors and development banks.

Government block grants generally provide core funding for research organizations, allowing them to focus on their strategic agenda of "public good" technology development. Safeguarding this funding requires vigilant attention to public relations and public awareness, effective service to clients, maintenance of political support for research, and establishment of and delivery on performance contracts. Indeed, skills in these areas should be a major criterion for selection of top research managers.

Competitive research contracts or grants from government or nongovernmental sources can add substantially to operating funds. Research staff need incentives to compete for these additional funds, as well as training in research proposal formulation. However, care should be taken to ensure that competitive funding does not divert the organization from its core research agenda, that quality products and services continue to be delivered on schedule, and – as far as possible – that competitive funding also covers related overhead costs for the organization.

Joint public-private ventures and other private-sector collaboration can be organized in various ways to increase resources available for research and to facilitate dissemination of findings. Here again, care is needed so joint projects do not subsume or appear to subsume the organization's own core research agenda of producing public-good technologies.

Commercialization of research products, agricultural products, or nonagricultural goods and services can provide research organizations moderate amounts of operating funds. However, these financing options frequently sound more attractive than they really are. Scientists and research organizations are typically poorly suited for operating commercial enterprises or marketing new technologies. When they do attempt to commercialize research products, such as new varieties, genomes, and machinery, they require qualified commercial expertise and legal advice, which can bring considerable costs.

Farmer funding, usually through broad-based farmers' organizations or levies, can provide a relatively stable source of financing for research on commercial crops (although this funding does fluctuate with prices and production). Producers should have substantial ownership and say over the use of funds generated from levies and such funds should not be earmarked exclusively for research in a public research organization.

Endowments are rare but an ideal source of stable funding for research – especially for long-term research activities. Their main drawback is that they are difficult to establish and, if large, can make an organization so financially independent that it ignores client needs.

Donors and development banks generally provide funding on a time-bound, project basis and are important sources for large investments in physical facilities and human resource development. Funding from these sources can often be obtained only with approval of the country's ministry of finance. If used to cover recurrent costs or research program expansion, donor funding may lead to serious financial difficulties when the supported project ends. It is imperative that financial planning look beyond the termination date of any donor-funded effort and develop a strategy for sustainable financing. As with private-sector funding, donor financing can distort an organization's research agenda, shifting it toward donor priorities.

Planning efficient use of funding

Along with defining funding sources and levels, financial planning must also ensure efficient use of resources. This involves a number of strategic choices among different categories of expenditures.

Research program costs versus overhead costs are a problem for organizations that have expanded and then faced budget reductions. Costs of administration, maintaining offices and laboratories, and operating research stations may leave little budget for real research. Overhead costs above 30% of the total budget suggest the need to review ways to improve efficiency.

Salaries versus operating costs may provide the clearest evidence of an imbalance in an organization's budget, as salaries take the first claim on resources. If salary costs exceed 75% of a research organization's budget, it is likely that scientists lack sufficient operating funds. A related issue is whether salary levels are adequate (competitive) to retain top scientists. These two issues often conflict, however, with total salary costs too high and individual salaries too low.

Recurrent costs versus investments represent a trade-off that can be ignored in an annual budget but not in the long term. Recurrent costs to run an organization and its programs are an obvious priority, but administrators that focus only on these short-term problems lose sight of the investment costs that are essential for maintaining or expanding an organization's capacity for future research. Investment costs are "lumpy," requiring large expenditures at irregular intervals. Key investments are in buildings, equipment, roads, irrigation and drainage, and staff training. Research managers might ignore these investments, since they may be included in donor-funded projects that are "out of the hands" of organization administrators. In analyzing investment budgets, financial planning might start with an estimate of the organization's total investment in buildings, equipment, and human capacity, as well as the annual depreciation of these assets. Though governments do not depreciate assets on an accounting basis, this concept is still valid and, if annual investment budgets do not match or exceed the rate of depreciation, the organization is essentially decapitalizing and cannot be expected to maintain its productivity into the future.

Strict controls must be balanced with flexible financial procedures. Controls are essential to safeguard resources and reputation, but efficient execution of research often requires a fair degree of flexibility in financial management. Agricultural and natural resource research tends to be time sensitive, because of planting and harvesting seasons, and long term, because of growing cycles or the need for multi-year trials. If funds for trials are unavailable when they are needed, both time and sunk costs are lost. Researchers, too, need some flexibility in managing budgets to accommodate unforeseen costs and to ensure that funds are available on time. This may require effective forward planning, authority for program managers to borrow funds, authority to switch line items, and the maintenance of an emergency account.

Block grants versus competitive funding may be a strategic choice for large organizations with numerous programs and projects. Allocation of budgets to research units for their use on a program basis is straightforward and minimizes transaction costs. On the other hand, introducing a competitive system for allocating a portion of the organization's funds may stimulate innovation, reward productive researchers, and help shift the focus of work in desired directions.

Outsourcing versus in-house service provision is an important, though politically difficult, decision for strategic financial planning. Research organizations may be able to contract services (e.g., farm operations, survey work, publications, janitorial services, laboratory analyses) more cheaply than they can provide them in house. Furthermore, some research may be executed more cheaply and effectively by outside contractors, especially if such contractors are farmers' organizations or other technology users. This strategy can be extended by using a competitive grant or contract system to implement a portion of the organization's research agenda. If appropriate research capacity and complementary skills exist in universities or elsewhere, it makes little sense for an organization to carry out the work itself at higher cost.

Balancing funding and program requirements

After evaluating and maximizing fund-raising and ensuring efficient use of funds, financial planning faces the difficult task of reconciling resource availability with the organization's needs. Rarely will available funds exceed program needs, so balancing financing with the organization's research strategies and agendas usually leads managers to a number of options.

The first and most palatable option is *fund-raising* to cover shortfalls in financing, although this may not be the best alternative if it yields only a temporary fix or if it diverts the organization from its strategic plan. Concerted effort by the organization to market its capacities and achievements is essential in raising funds. A research organization that can demonstrate it is efficient, productive, and responsive to stakeholder needs will usually have little difficulty in funding priority programs.

Consolidation of research infrastructure is a second option, and one that can generate substantial cost savings. Most large research organizations control lands scattered across many research stations and substations several times the size required to carry out the planned research program. These properties may be costly to operate, lack critical mass of scientific staff, and be of little relevance to the main research agenda. Closing such stations, selling some facilities or turning them over to other agencies, moving research to farmers' fields and organizations, and developing collaborative arrangements with organizations such as universities may substantially reduce overhead costs with relatively little impact on programs. This may, however, risk significant political costs and opposition, including resistance from staff affected by the consolidation.

Eliminating low-priority or unproductive programs is an essential part of financial planning. Funding a few programs adequately is always preferable to spreading resources too thinly over many programs. It is usually easier to redirect operating costs to priority programs than to reassign staff who may have to be retrained or terminated.

Rightsizing staffing levels is perhaps the most painful means of adjusting the organization size to fit available funding, but is often necessary. This requires attention to the numbers of scientists that can be adequately supported over the long term, ratios of technical support staff to scientists, and numbers of administrative staff. Effective personnel systems for annual performance reviews are essential for identifying unproductive staff. Also, the rapid spread of information technology now offers ways of reducing numbers of administrative staff.

Finally, *efficient financial systems* are essential to good financial planning and to balancing resources and program. Budgets and accounting systems should provide timely financial reports that reflect plans and expenditures by program and project, by scientist, and by category of expenditure. Other management information systems can be developed to track the performance and impact of research, serving as a basis for evaluating efficiency and planning future financing.

Relevance for agricultural research

Financial planning is an essential element of a research organization's strategic plan. However, in practice, many such organizations live with a high degree of uncertainty with regard to funding. Government policies and budgets may change with new economic conditions, governments, or government ministers. A long-term financial plan helps an organization to recognize potential problems early and take actions to ameliorate financial problems. Well managed organizations are better able to maintain their funding base than poorly managed ones.

Having a financial plan also aids fund-raising. Stakeholders can be presented with a comprehensive picture of the organization's needs and plans for sustainable financing and details of what their financing will buy. Adequate plans may also help fund-raisers match sources of financing with particular needs. For example, a donor might be approached to fund lumpy investment costs or a farmers' organization might be asked to fund certain operating costs.

Finally, the discipline of long-term financial planning helps managers look beyond the current financial situation, which may in fact be an aberration because of the organization being flush with donor funds or starved due to structural adjustment. A long-term perspective provides a more accurate picture of the organization's real financial health. Regrettably, it often must lead to hard decisions on balancing resources and programs.

Examples

Few research organizations undertake long-term financial planning; so there are few documented examples of good practice. However, many public-sector research organizations have employed various financial instruments to meet medium-term goals of research execution.

Brazil

The national agricultural research corporation of Brazil, "Embrapa," has been relatively successful in maintaining stable financing. Embrapa has developed an admirable record of research achievements and efficient management that has engendered confidence in its ability to effectively utilize the funds that it raises. To support performance and quality-oriented research management, Embrapa developed a evaluation system that includes client feedback for its different research centers. Evaluation results are used as one consideration in allocating funding to its organizations and programs. Based on solid research results and keen public awareness, Embrapa has been able to foster broad political support for agricultural research among urban consumers. In part, this was accomplished by a very effective advertising campaign on television that helped stakeholders to see and appreciate Embrapa's impact. Embrapa managers have also maintained effective communication with the finance ministry on the payoffs to investment in agricultural research and the public-good nature of much of the work it carries out. Although Embrapa receives most of its funds as block grants, it recently sought to diversify funding by seeking new partners in the private sector and introducing competitive funding for selected programs to encourage linkages with other public, nongovernmental, and private organizations.

Australia

Since 1985, Australia has sought to expand farmer financing for agricultural research and increase efficiency in use of research funding. The government established 16 research and development corporations (RDCs), most with a commodity focus. The government cofinances research in these corporations by matching funds provided by farmers and industry up to 0.5% of the gross value of production of the commodity. The research corporations, which account for approximately 30% of Australia's agricultural research investment and a much higher share of operating costs, contract out all of their research, most of it through competitive bidding. The administrative costs of running the corporations has been kept to less than 5% of total research funds.

China

Commercial operations rarely provide a reliable or optimal source of funding for public research organizations. Yet economic reforms in China led the government to encourage research organizations to generate their own funding. In response, organizations started a wide range of income-generating activities: agricultural production, research-related endeavors, and even non-agricultural occupations. Unfortunately, government financing tended to be reduced in proportion to the resources generated by these commercial operations. After considering the cost of the commercial operations, the net income generated for agricultural research in most cases appears to have been negative.

Recommended reading

Tabor, S., W. Janssen, and H. Bruneau. 1997. *Financing Agricultural Research: A Sourcebook.* The Hague: International Service for National Agricultural Research.
This book provides in-depth coverage of the topics touched on here, including chapters on capital investment, operating costs, donor assistance, private funding, and financial systems.

Beynon, J. et. al. 1998. *Financing the Future: Options for Agricultural Research and Extension in Sub-Saharan Africa.* Oxford: Oxford Policy Management.
A good overview of issues in funding research, including case studies of experiences in Kenya and Zimbabwe.

Alston, J. M., P. G. Pardey, and V. K. Smith (eds.). forthcoming. *Paying for Agricultural Productivity.* Baltimore: Johns Hopkins University Press.
An in-depth treatment of financing principles for agricultural research with case studies from a number of countries.

Chapter 14
Planning Training

Edwin Brush

Training planning is the process of designating what knowledge, skills, and attitudes are to be developed by staff of a research organization. The relevance of training planning is linked to two challenges that agricultural research systems face today. First is the scarce supply of staff suitable for training in degree programs. Second, and of particular relevance for nondegree program training, is the need to increase training's impact on workplace performance. The planning process outlined here can be used for both degree and nondegree programs. It has four sequential sets of activities: preparing for planning, analyzing training needs, budgeting and scheduling training programs, and following up on the training plan.

What is planning training?

Although the definition of "training" varies among organizations, this chapter takes it to mean the purposeful development of employees' knowledge, skills, and attitudes (KSAs) so these can be applied in current or future jobs (Abe 1990, Patrick 1992). Training planning is the process of stipulating which KSAs to develop under the auspices of an organization and how such development should take place. Purposeful development can be more or less formal, ranging, for example, from classroom instruction to coaching during certain job assignments (Brush 1993). Here we concentrate on the more formal sense, that is, activities involving explicit instruction. Agricultural research organizations generally use two types of formal training. One is degree training, which involves comprehensive intellectual growth over a long term (measured in months or years), typically through accredited programs resulting in, for example, a degree or diploma. The other is nondegree training, that is, short-term development (usually measured in days or weeks) of particular KSAs needed to perform a job or function. The various programs that make up this type of training are usually unaccredited (e.g., short courses, workshops, and on-the-job training).

Training planning in an agricultural research organization, then, is the process of choosing goals for degree and nondegree training and specifying what, when, where, and how training programs should be undertaken by which staff. Its purpose is to provide targets for investments in human resource development during a specific period. Planning can make long-range, strategic projections encompassing many training programs over several years; or it can be more operational, covering training in a single year or a single training program (see also Wentling 1993). The focus here is on long-range planning. Given the long-term nature of many degree programs (a PhD may take four to five years to complete), training plans are often obliged to predict KSA needs as far as five years in the future.

Doing training planning

Specifying KSAs to purposefully develop in an organization is a complex challenge for which there is no single commonly accepted "correct" approach. Here, we discuss four sequential sets of activities in training planning: preparing for planning, analyzing training needs, budgeting and scheduling training programs, and following up on the training plan (Abe 1990). Rather than prescribe activities, we suggest options for agricultural research organizations to consider in light of their local situation. Organizations are encouraged to modify activities discussed here to better fit their own context.

The first set of activities involves preparing for the planning process. An initial step is to determine the range of the plan (three to five years is typical) and whether such planning should be a regular or occasional exercise. An ongoing planning effort in which training goals, schedules, and budgets are updated during the planning period, may be required if significant investments are being made in training. The staff time needed for recurrent planning exercises will be justified in such cases. Moreover, repetition can improve the effectiveness and efficiency of planning.

Assigning responsibility for leading the process is a crucial initial step. Some research organizations have a training officer who can take charge; in others the head of the personnel department can lead. This manager will need to determine what level of participation is desired and feasible in the planning process. Participation may be limited to a top-level task force of managers who provide input from an organization-wide perspective on, among other things, research strategies and priorities, performance targets, and technological changes that influence the KSAs needed. An alternative is to broaden participation to include scientists and support staff. Expanded participation may contribute more to fine-tuning the analysis of training needs than to budgeting and scheduling. An advantage of limiting participation may be reduced resource requirements (i.e., less time and money). An advantage of expanded participation may be access to a wide body of information.

The second set of activities comprises the analysis of training needs – identification of KSAs presently or expected to be deficient in the organization. It includes designing analytic procedures, collecting information, and carrying out the analysis (Salem 1986). An initial step is to select questions about the organization being analyzed. Among others, questions in five sequential categories are usually examined (Laird 1985):

1. What are the KSAs in the current organization according to program, function, level, discipline, and qualification of existing staff?
2. What changes are anticipated and what KSAs will be required at the end of the planning period (e.g., retirements, changes in the research strategy or staffing plans, KSAs linked with desired improvements in performance of individuals, programs, functions, or the organization)?
3. What is the gap between the current supply of KSAs and future demands? Which gaps are most critical? What are constraints to applying KSAs to improve performance at different levels of the organization?
4. Which of the critical gaps, if any, can be filled by non-training solutions (e.g., relocating staff, redesigning jobs, contracting services, improving management processes, promulgating guidelines, and enhancing incentives)?
5. Which gaps should be filled by degree and nondegree training programs? How many staff from which areas should be trained in which KSAs at which level? Which training programs should be given priority? How can the success of gap-filling be monitored?

The next activity in this set is to identify staff and documents to help answer the questions posed above. Staff sources may include supervisors and others. In small research organizations all staff may be asked for information; in a large agricultural research institute, sampling procedures may be necessary (stratified samples are often used to account for staff differences according to, among others, position, education, and gender). Documents may include reports from a management information system, the organization's strategy, research priorities and plans, staffing plans and policies (relating to positions in various units and the capacity to provide staff to fill those positions), the training policy, analyses of jobs and organizational constraints, and staff performance evaluations. Not all sources will be available when planning is initiated. While others are useful, the most crucial sources are managers' input, the organization's strategy, and research priorities. Once available sources are identified, they are tapped for information on selected questions through surveys, interviews, group sessions, workshops, search and retrieval of documents, and so forth. Collected information is then analyzed to formulate answers as to which degree and nondegree programs to implement.

The third set of activities involves budgeting and scheduling programs to fill the needs identified through the previous analysis. An essential input is budget information from funding sources for training. This may concern national grant schemes, bilateral programs (also exchange programs) or multilateral programs. Budget information is used to specify the funding source and to estimate research needs for each program. Estimates are based on cost standards, for example, per-year costs for degree programs and per-participant costs for short courses. Typically, the output is a plan that shows training programs to undertake annually during the planning period. This plan lists training by topic for specific staff on a year-by-year basis. Probable locations for degree programs (whether national or foreign) and nondegree programs (whether internal or external) are located based on the research organization's experience and donor requirements.

The fourth set of activities involves following up the training plan. Follow-up includes operational planning for specific programs to be undertaken during the year (Wentling 1993) and the monitoring and evaluation of programs already implemented (Mabeza 1993). Decisions by managers and donors to release staff and funds for training are based, in part, on training plans; however, regular reports on training activities and impact help maintain management and donor commitment. This leads to a need for regular reporting on programs that have been implemented. It also provides opportunities to bring the plan up to date. One strategy is to use the rolling-plan approach, which combines regular monitoring of training events with an annual review of needs and updating of the plan. An annual review and plan update may require less effort for data collection, analysis, and preparation than was required in the initial year of the planning period.

Relevance for agricultural research

Traditionally, training has been an important means to strengthen agricultural research capacity and a prominent feature in many multilateral and bilateral projects. However, today, continued investment in training is being challenged on two fronts. First, research organizations face an era of fiscal austerity, which makes it difficult to sustain established patterns of training. On the second front, training has lost some of its appeal as a means to strengthen research capacity, since some past training has failed to produce measurable results in terms of improved performance of research organizations. This may be because training has been poorly planned, or because the staff member trained was not placed in a position where she or he could enhance performance, or because evaluations of training impact were scanty or lacking. The principal relevance of training planning is to enable research organizations and donors to face these challenges and make their investments in training more productive.

The first challenge evolved, in part, from the success of training during the 1970s and 1980s. A priority then was to increase the numbers of national agricultural scientists working in research organizations in developing countries. The results were impressive: numbers more than doubled between 1971 and 1985. Growth often followed a pattern in which new staff were hired then trained in degree programs. In many developing-country agricultural research organizations today, more than half of the scientists hold an advanced degree (MSc or PhD). In the stark funding scenario of the 1990s, concern shifted from managing growth to managing austerity. Many agricultural research organizations were unable to hire staff even to replace departing scientists. Low growth, coupled with an already highly trained staff complement, have reduced the supply of staff for degree training (Brush 1993).

Nonetheless, training needs persist – alongside the need for financial support from donors for degree training. Demand for KSAs encompasses new disciplines such as biotechnology and natural resource management, as well as expansions in traditional research programs. Given the scarce supply of staff for degree programs, training planning is particularly relevant to help ensure that investments in training effectively match the scarce supply of staff with research priorities. Planning helps research organizations contend better with human resource constraints. It enables them to manage training from a strategic, organization-wide perspective, rather than from more limited perspectives such as those of single programs or functions.

The second challenge for planning involves the failure of training to adequately demonstrate an impact on performance, especially at the level of the organization, for example, on a research program or organizational unit (Mabeza 1993). This challenge is particularly important for nondegree training. Many factors contribute, including, among others, attrition and transfer of trained staff, lack of institutional support for implementing lessons learned in training, insufficient fit between training and institutional objectives, and financial constraints (e.g., poor incentives and inadequate operating funds). Certainly, training has provided significant benefits to individuals, giving them, in addition to enhanced KSAs, qualifications for promotion, financial gains from travel, and valued personal and professional contacts. The challenge is to shift the benefits from the individual to the organizational level. This challenge is related to the challenge on the first front: when staff for degree training is scarce, appeals may be made to redirect funds from degree to nondegree training. To answer this call, the impact of degree training must be demonstrated beyond the individual level so that such investments are sustained.

Planning increases the probability that training investments will result in improved performance at the organizational and individual levels. Results can be expected most where training is planned explicitly to change organizational behavior. For example, the planning process might

- identify the functions in which training can most effective, since constraints have been recognized and can be removed
- show that a certain management procedure should be designed and approved by the organization prior to training in this procedure

- specify that training for staff in a particular unit should go ahead because supervisors in that unit have agreed to encourage their staff to apply KSAs learned in training
- settle on procedures for selecting staff for training and for following up afterwards in order to enhance opportunities for institutional development

Ways to shift training benefits to the organizational level need to be addressed in operational planning of individual programs. Moreover, agreement on how training can best be targeted for organization-wide impact should be established during the overall planning process.

Examples

Training planning has been a key aspect in developing agricultural research capacities for many years, often linked with multilateral and bilateral projects. ISNAR has collaborated in a number of planning efforts. Two recent examples illustrate the processes – agricultural research organizations in Kenya and Bhutan.

Kenya

The Kenya Agricultural Research Institute (KARI) undertook with ISNAR to develop a five-year "training master plan" for the period 1998–2002. A training needs assessment was a major input to the plan (ISNAR 1997). An assessment exercise was carried out through a series of workshops that included of about 15% of KARI's managers, scientists, and support staff. The exercise identified and prioritized gaps in technical and management knowledge and skills. These became the basis for planning nondegree programs. The training master plan included the history of training at KARI, staffing and training policies, the plan's rationale, and a summary of the training needs assessment results. Degree and nondegree programs were proposed year by year with staff targets, budget estimates, and donors where known. Finally, guidelines were provided for implementing the plan.

Bhutan

In Bhutan, the national agricultural research system was reorganized in 1992 (ISNAR 1992). A new research strategy and plan for the renewable natural resources sector called for new technical and management KSAs. Staff abilities were compared with needs for implementing the new strategy. The comparison revealed gaps that could be filled by training staff and recruiting and training new staff. Bhutan prepared a five-year plan of top-priority training needs, including degree and nondegree programs. The plan also indicated other needs that were to be subsequently tackled.

References

Abe, L. O. 1990. Some Aspects of Training in National Agricultural Research Systems. Training Series, Human Resource Management No. 3. The Hague: International Service for National Agricultural Research.

Brush, E. G. 1993. Human resources for agricultural research: Issues for the 1990s. *Public Administration and Development*, 13 (3): 295–305.

ISNAR. 1997. Framework and Methodology on Training Needs and Organizational Constraints Assessment in Kenya. The Hague: International Service for National Agricultural Research.

ISNAR. 1992. Review of the Bhutan National Research System: The Renewable Natural Resource Sector. Country Report No. R53. The Hague: International Service for National Agricultural Research.

Laird, D. 1985. Approaches to Training and Development. Reading, Massachusetts: Addison-Wesley Publishing Company.

Mabeza, H. 1993. Training evaluation. In *Monitoring and Evaluating Agricultural Research: A Sourcebook* edited by D. Horton et al. Wallingford, UK: CAB International.

Patrick, J. 1992. Training: Research and Practice. London: Academic Press.

Salem, C. I. 1986. Assessing Trainability: A Workbook. Economic Development Institute (EDI) Training Materials, Course Note 670/030. Washington, D.C.: World Bank.

Wentling, T. 1993. Planning for Effective Training: A Guide to Curriculum Development. Rome: Food and Agriculture Organization of the United Nations.

Part III
Agricultural Research Planning as an Institutional Process

Govert Gijsbers

The nature of the planning process

Traditional models have assumed the planning process to be linear, rational, and comprehensive. But most practitioners experience planning as more messy, intuitive, and patchy than they had initially expected (De Wit and Meyer 1998). To organize planning effectively, research managers must be aware of the reality of planning processes, instead of basing their actions on common misconceptions.

First, planning is often thought to consist of a linear sequence of steps: analysis, planning, and implementation. Yet in most cases the process is muddled, with analysis, planning, and implementation taking place simultaneously. New plans may be formulated, approved, and perhaps even implemented, while "old" plans continue operations. Budgets may be cut in response to changes in the country's overall fiscal situation, irrespective of what research was planned. Moreover, funding agencies operate on a variety of planning and funding cycles; research organizations can separate themselves from these only at great cost. In short, the world does not simply stop for an organization to contemplate its current situation and formulate its plans for the future (Mintzberg 1994).

Second, most planning processes have both rational and intuitive elements. Part of the process might rely on quantitative tools and methods: planners develop projections, estimate costs and benefits of different strategies, propose different funding scenarios, match funding with priorities, and calculate financial and human resource requirements for various levels of organizational activity. Intuition is required to define a strategic vision and direction, to judge opportunities and threats, and to assess and agree on which fields are particularly promising areas of new endeavor.

Third, planning may aim to be comprehensive, covering an entire organization. But it normally leads to incremental change and shared learning rather than a redesign of the organization as a whole. Most organizations are highly complex; it is impossible to revamp them completely without major upheavals. Change should be conceived as taking place in small steps; testing what is feasible, overcoming resistance, orienting different departments in the same direction, and developing shared understanding of problems.

Organizing implementation-oriented planning

Ultimately, the quality of a research plan depends on the extent to which it is implemented and, therefore, to the degree that implementation improves the performance of the organization. A number of factors may contribute to success in plan implementation.

Actors in the planning process

No planning succeeds without leadership that supports it and gives it legitimacy (Bryson 1995). Key decision makers must be involved from start to finish. Someone has to take on the roles of "sponsor" and "champion." The sponsor may be an individual such as the director, a committee within the organization, or an outside body such as a parent organization, and may or may not be the same unit or person as the process "champion." A champion is needed to guide the process, to motivate people, ensure that tasks are completed, meetings held, feedback obtained, and a plan document delivered. Sponsors provide legitimacy, while champions provide energy and commitment to sustain the process. The champion focuses on facilitating the planning process, having no preconceived ideas about desirable outcomes.

The process champion leads the selection of participants in the planning process and assigns responsibilities to internal and external stakeholders. Two categories of stakeholders are often distinguished. Stakeholders in the wide sense are those groups that can affect the performance or are affected by the performance of the organization. Stakeholders in the narrow sense are identifiable groups on which the organization depends for its survival. For agricultural research, the first category includes other research organizations, extension services, environmental groups, public interest groups, trade associations, government agencies, as well as the stakeholders in the narrow sense, such as employees, clients, and financiers. Deciding who should be involved in each step of the process, what the responsibilities of each stakeholder should be, and what information they should receive and provide are key issues in this respect. The broader the scope of the planning effort, the larger and more diverse should be the stakeholder group participating. At the same time, broader participation adds significantly to the complexity of the planning exercise.

Integration of planning with other functions

Plans can only be implemented when they are embedded in the organizational processes, functions, and structure. Planning must be integrated with administrative cycles (financial years and government and funding agency planning cycles). Awareness and, where possible, synchronization of the different internal and external planning, political, and administrative cycles are therefore important conditions for effective planning and implementation. If the different cycles are incompatible or cannot be synchronized, the scope for effective planning is reduced and the organization will need to learn to live with considerable uncertainty. To deal with uncertainty explicitly in planning and implementation, some organizations have adopted "rolling plans" that are adjusted periodically, in the margins or in more fundamental ways, depending on the nature and direction of internal and external change.

The ability to adjust planning and implementation processes depends on the availability of information about the continued relevance and feasibility of the current plan. Internal changes that may invalidate a plan under implementation include changes in resource availability and in leadership. External changes are those that affect the stakeholders to such an extent as to necessitate adjustments within the organization serving them. Examples are research breakthroughs at other organizations and changes in government policies and extension organizations. Monitoring and periodic evaluation of the plan and its implementation, using indicators or "milestones," provides decision makers with information that allows them to change course midway.

To ensure that relevant information is in fact provided, a number of monitoring and evaluation mechanisms should be in place. These range from informal, low-cost mechanisms to formalized, costly ones. A mix of informal (e.g., project discussions and trip reports) and formal mechanisms (e.g., annual program reviews and databases using quantitative sets of indicators) is arguably the best way to monitor ongoing research. To integrate planning, monitoring, and evaluation into the life of the organization, annual planning meetings, internal and external reviews, and annual reports can be institutionalized.

Integration with organizational structure

Many large research organizations have established permanent planning units, planning cells, or units for planning, monitoring, and evaluation. Often, these units were set up to deal with donor-funded projects and subsequently were expanded to cover all research activities. While a perceived need for formalization of planning is the rationale for maintaining such units, the question may still be asked whether planning is best undertaken by such units, or in a more ad hoc manner by flexible teams from different departments. In practice, planning units in agricultural research organizations are often overwhelmed with the paperwork that is typical of public-sector bureaucracies and, as a result, cannot pay much attention to real planning.

Whether a planning unit has significant influence on the substance and process of research planning depends to a large extent on its place in the organization and on the support it receives from senior leaders. There is general consensus that a planning unit should be placed close to the chief executive officer of the organization and should function as a part of top management. There is less agreement on who should staff the planning unit: whether positions in the unit should be permanent or if staff should rotate between technical departments and the planning unit. Continuity is important, but the planning unit should not become isolated from the rest of the organization. This risk is particularly great when planning unit staff are mainly social scientists, while the rest of the research organization is dominated by researchers from the life sciences.

One way to deal with the issue of staffing is to keep the planning unit relatively small in size and rely on technical department staff to carry out specific planning exercises with assistance of the planning unit. There should be some senior, well respected staff in the planning unit, however, to ensure cooperation from the line departments. And the planning unit should house a mix of disciplines. With the nature of planning evolving towards more participatory and consultative approaches, the planning unit's facilitating role is increasing. Skills in facilitation are urgently needed. Other vital areas of expertise relate to the emerging fields and challenges that are shaping the environment in which agricultural research takes place. Examples are intellectual property rights, environment and food safety, assessment of new technologies, and trade policies and patterns.

Identification of clear and doable priorities

Selecting a realistic and coherent portfolio of priority projects for each program is a key task in many planning exercises. There are several approaches, tools, and techniques for planning and priority setting in agricultural research organizations. The methodology chosen depends on the type of organization, the type of research, and the analytical capabilities available. The number and type of priority projects that result from the planning exercise should be in line with the likely availability of financial and staff resources. Some organizations identify three project portfolios, consistent with "optimistic," "pessimistic," and "most likely" levels of funding. The practice in many organizations, however, is to grossly overbudget in the hope of receiving at least a small portion of what was requested.

To implement a plan it is essential that priorities can actually be translated into changes in resource allocation. New priorities can be operationalized by making additional funding available. But often it requires low-priority activities to be discontinued to make place for higher priority endeavors. Such changes often incur considerable resistance from those affected and usually are difficult to achieve without senior management support.

The chapters

The first chapter in this part on institutional aspects of planning focuses on the professional staff and specialized units that support planning processes. The *roles of planners and planning units* are changing as a result of organizational changes in many agricultural research organizations. The chapter discusses ongoing changes in the institutional landscape, with particular emphasis on the shift from traditional, public-sector bureaucratic traditions to more open and flexible network-type institutions.

The chapter on *enhancing participation* looks beyond the role of planners and discusses how a wider variety of stakeholders may be effectively involved in the research planning process. It discusses the benefits of participation, particularly in terms of improved relevance, representativeness, equity, and ownership.

The chapter on *priority setting* explores the links between planning processes and priority-setting exercises. Different levels of priority setting (national, institute, program) are discussed, along with decisions that need to be made when doing priority-setting work. These include who to involve, how much to spend, what methods to try, and how to define the range of research alternatives from which to select priorities.

Resources are allocated through the budgetary process, which plays a major role in short-term planning. While scarcity of staff and equipment may be problematic, it is usually lack of funding in the right amounts and at the right times that causes havoc in implementation of agricultural research projects and experiments. The chapter on *planning and budgeting* highlights the importance of budgeting as a tool for translating plans into action.

Plan implementation has long been ignored, because implementation has been assumed to be "automatic" once a plan was adopted. The chapter on implementation discusses a number of ways in which plans can and do go wrong in the implementation phase, and what measures can be taken to improve the chances of successful implementation.

Finally, the chapter on *linking planning to monitoring and evaluation,* elaborates the importance of establishing an integrated planning, monitoring, and evaluation system that feeds information collected during and after implementation back into the planning cycle.

References

Bryson, J. M. 1995. Strategic Planning for Public and Nonprofit Organizations: A Guide to Strengthening and Sustaining Organizational Achievement. San Francisco: Jossey-Bass.

Mintzberg, H. 1994. The Rise and Fall of Strategic Planning. New York: Prentice Hall.

De Wit, B. and R. Meyer. 1998. Strategy: Process, Content, Context: An International Perspective. London: International Thomson Business Press.

Chapter 15
Roles of Planners and Planning

José de Souza Silva

Institutional and organizational changes have a profound impact on the roles of planners and planning. In the past, the dominant organizational model was the "rational organization," with efficiency, predictability, quantification, and control as its guiding principles. The traditional model saw planning as a bureaucratic activity in which planners were responsible for data collection and analysis and the formulation of programs and projects. This model is now being replaced by a "network model" for organizing agricultural research. The network model emphasizes collaboration, consultation, individual responsibility, and flexibility. The production of new knowledge in networks is closely linked to application of that knowledge. It is transdisciplinary in nature, includes a variety of actors and accountabilities, and uses a broad concept of quality control. In organizations following the network model, planning emphasizes strategic thinking. Planners support management's exploration of alternative research options, rather than defining a single, best alternative.

What is the role of planning and planners?

Turbulence and complexity increasingly characterize the context in which agricultural research is conducted. New developments in information technology and the life sciences are reshaping the way the agricultural sector and agricultural research are organized and managed. Agriculture is becoming knowledge-based rather than resource-based, and knowledge is becoming the main driver of productivity throughout the agricultural and agroindustrial sectors. Globalization is continuing to affect agriculture research organizations (see Tabor, this volume), and agricultural research agendas are gaining complexity. The emerging network organizational model is better suited to deal with complexity and uncertainty than the traditional rational model. Yet organizational transformation towards the network model will have a profound impact on the roles of planners and planning units.

The rational organization

The rational organizational model is based on the classic "scientific management" methods of Frederick Taylor and Henry Ford. In this model, individuals are supposed to follow rules and structures that steer them towards optimal work procedures, behavior, and solutions to problems. Individuals do not take initiative or design their own objectives and activities. Rather, the model is characterized by efficiency, predictability, quantification, and control.

The efficiency dimension of planning. In rational organizations, planning units are designed to efficiently handle a great deal of paperwork and to process a massive amount of data in their program-formulation duties. Planners are thus seen as data-gatherers and processors, and planning units known as repositories of documents and databases. As specialists in the efficient use of resources, planners and planning units also contribute to the creation and institutionalization of methods and rules to establish "best practices" at research management and plan implementation levels. Now evident, however, is that the resulting ever-growing number of norms and procedures is likely to develop into an inflexible bureaucratic obstacle.

The predictability dimension of planning. Predictability implies stability. This assumption of a stable world, only rarely disturbed by temporary conflicts, has shaped the views of most managers and planners. Planners and planning units assume that the world will remain still so that plans may be implemented as programmed. Moreover, organizations expect planners and planning units themselves to be predictable: plans, programs, and projects are supposed to be delivered regularly and in neat packages. Researchers also assume that planners and planning units will produce guiding frameworks each year to support the process of programming next year's activities.

By assuming stability, planners convey the belief that the future will be very much like the past or present. Historical data series then become the most reliable information source for predicting the future. With these in hand, planners consider themselves sufficiently informed to set a research agenda. In this scenario, participation of stakeholders is unnecessary for making research policy and programming decisions and setting priorities. It may disturb the planning process. Sequential planning of agricultural research seems reasonable, in a linear fashion from analysis to decision and action. Researchers are supposedly able to predict accurately how much time their research project will need to produce the expected results. This expectation often leads scientists to look for simple research problems that are amenable to experimentation under controlled conditions.

The quantification dimension of planning. Rational processes rely heavily on quantifiable attributes and aspects of reality. To demonstrate organizational performance, planners and planning units have emphasized hard data over softer types of information, quantitative aggregates over qualitative phenomena, measurable facts over interpretative "soft" facts, and quantifiable objective goals over hard-to-measure subjective ones. The rational model's emphasis on quantification has had considerable impact on agricultural research planning. Agricultural scientists have emphasized volume of production at the expense of sustainability, competitiveness, product diversification, safety, and quality. Quantification in planning demands that monitoring and evaluation models incorporate primarily data that can be measured. Thus, critical environmental, social, political, ideological, ethical, and institutional factors have been systematically excluded from the planning process.

The control dimension of planning. In most organizations, planning is carried out by planners located in formalized planning units. This implies a centralization that separates planning from implementation and gives planners a great deal of control, leaving little room for incorporating knowledge and judgments of other key actors outside the planning unit and outside the organization. Planners and planning units have also held great sway over budgeting – to the point that many view budgeting as synonymous with planning. As a result, planning has been viewed as technical, neutral, and value-free, an activity carried out in an objective fashion.

Nonetheless, there is often conflict around the process of knowledge and technology development. Planning is therefore not a neutral process. The control dimension of planning may be further emphasized by the use of computerized information systems, particularly if these are designed in a top-down manner that does not empower its users.

Implications of the rational organizational model

The dominance of the rational organizational model in planning has had a number of implications. First, the drive for increased efficiency leads to an exaggerated effort to improve processes that are internal to research without a correspondent effort to improve the impact of the knowledge and technology resulting from it. Second, emphasis on prediction leads to supply-oriented research models that neglect congruence between research results and demands of stakeholders in the agrofood and agroindustrial chains. Third, emphasis on quantification has led to neglect of other aspects that are important but difficult to measure. Fourth, the drive for control prevents a grounding of the planning culture: in most agricultural research organizations, managers and researchers view planning as a bureaucratic activity. Institutions are obligated to plan, but the relevance and contribution of planning is seen to stop at budgeting.

The network model

Network organizations follow a different logic, emphasizing application-oriented knowledge production by teams of diverse individuals from a variety of disciplines. These actors are held accountable by stakeholders and apply a broad approach to quality control. Networks are open structures that rely on horizontal rather than vertical organizing principles. In egalitarian networks there are no superior-subordinate relationships among participants. Networks are asymmetrical. But each node of the network can hardly survive by itself; neither can it impose dictates on the other participants. The organizing principles of networks differ from those of traditional organizations in that (i) participants are included because of their interests in, and ability to contribute to, network objectives, (ii) network members are "loosely coupled" and participate in system activities rather voluntarily, (iii) actions and decisions revolve around a broad vision or purpose and a set of goals and objectives that reflect the interests of network members, and (iv) there is usually no central source of power in a network although there may be a need for a strategic central node to assume functions that are critical for the sustainability of the network. The network model helps share uncertainty and mobilize strengths among participants while reducing weaknesses at the level of the individual network members.

A network consists of connected nodes that are both autonomous and interdependent. Nodes, which are participating organizations or individuals, may be a part of other networks, and therefore of other systems aimed at other goals. The performance of a network depends on its ability to facilitate communication between its components and on the extent to which there is sharing of interests between the network's goal and the goals of participants.

The networking mode does not imply the disappearance of organizations as such. Rather, most organizations would function within different networks, and modern organizations often operate internally in a network mode. At the same time, the horizontally oriented network mode of organizing activities and interactions may gradually replace the vertically oriented rational mode of organizing, even in large corporation or public-sector organizations.

Michael Gibbons and colleagues (1994) have described the "new mode of knowledge production" as flexible, because of its external orientation towards stakeholders. Flexibility is seen in five characteristics of this emerging mode of knowledge production: (i) context of application, (ii) transdisciplinary effort, (iii) large heterogeneity of actors and diversity of organizations, (iv) increased social accountability, and (v) broad quality control. The emerging mode of knowledge production is more likely to take root in open-ended networks than in traditional, closed rational organizations. This new mode of knowledge production has a number of implications for planning.

Planning and application-oriented knowledge production. In the new network mode, the production of knowledge is directly linked to its application. Usefulness and problem-solving are central concerns. The dynamics of the organization's context require planners to build uncertainty into the analysis and to design ways that planning can help the organization respond to changes in its environment. Application-oriented actions are not amenable to excessive formalization. Deregulation of most vertically structured rules is necessary, since top-down control is impossible. Flexibility of management and planning in applied agricultural research is the rule, and centralized planning processes and supply-oriented research models are replaced by demand-driven approaches. Planning is no longer seen as separate from implementation.

Impact of transdisciplinarity on planning. An external orientation that takes into account the broader context of research often results in a more complex definition of the problem to be solved. Diagnosis, analysis, and action associated with complex problems require a transdisciplinary effort. Transdisciplinarity is the most important form of knowledge production in the new mode. It implies going beyond disciplinary boundaries to incorporate the knowledge and judgment of other experts in a context-related analysis of problems and the solutions proposed. Characterized by a constant and deliberate flow of information between the theoretical and the applied dimensions of research, transdisciplinarity arises only if research is based on a common theoretical understanding achieved through a collectively constructed conceptual framework.

Transdisciplinary projects pose demands that change traditional planning. First, planning is not just for planners: all actors participating in the research effort must be involved in the process to fully understand the complexity of their collective task and cohere themselves into a team. Second, since team outcomes are the product of consensus, planning becomes a process of permanent negotiation. Third, creativity in planning becomes a team-dependent phenomenon, not the product of gifted individuals. Finally, in order to be active participants in the planning process, all social actors have to be trained in planning skills. Thinking cannot be separated from action in the network model. Moreover, flexible planning works better under the network model, especially if members of multidisciplinary teams belong to diverse organizations that are independent of one another.

Diversity of actors. Diversity or heterogeneity of actors under the network model refers to the differences in knowledge, experience, and skills that actors bring to the process of knowledge production. Managing heterogeneity requires talent for identifying and managing professional skills, as well as for brokering knowledge.

Individuals participating in planning will usually come from various organizations. Such organizational diversity increases the number of potential sites for knowledge production and implies the need to link the different sites through communication networks. Cooperation and competition, resistance and conflict are all common features of operating in the network context. Since conditions for prediction and control are almost nonexistent in heterogeneous networks, traditional planning needs to be transformed in order to facilitate interdisciplinary and interinstitutional negotiations for building understanding, consensus, and commitment. To allow satisfactory participation in such a diverse group of social actors, qualitative and participatory action-research methods become increasingly relevant.

Social accountability. The growing awareness that science and technology play an important role in development has led to social groups' growing interest in influencing technological change. The process of knowledge production should involve those interested social actors in order to ensure accountability. Technological solutions to problems may touch upon the values and preferences of social actors that were traditionally seen as operating outside of the science and technology system. Such actors now become active agents in the definition and solution of problems, as well as in the evaluation of organizational performance and research results. Research issues can no longer be answered in scientific and technical terms alone. Social scientists increasingly participate in context-oriented, transdisciplinary projects. Agricultural science and technology organizations may have to create fora for broadening public influence and social control over research projects. Specialized planners, managers, and researchers will be unable to establish their leadership through the use of formal authority alone; intellectual, managerial, and technical competence will garner allegiance in a more democratic, network-like environment.

Planning and quality control. In the traditional mode of knowledge production, quality control is left to peer review. The new mode of knowledge production, however, demands a broader process of evaluation that incorporates additional criteria from the context of application, including social, economic, and political criteria. In application-oriented quality control processes there are different definitions of what constitutes valid knowledge, successful technological solutions, and useful research results. Planning, monitoring, and evaluation systems are reconfigured to incorporate a broader criteria matrix for supporting the requirements of the new quality control systems. Sooner or later, *sui generis* models of evaluation may need to be created to evaluate networks. These models are indispensable for incorporating a diverse range of criteria, as well as all the subjective, qualitative dimensions of context-oriented, process-dependent, transdisciplinary research projects. Key network members will demand inclusion in any network-related evaluation.

Relevance for agricultural research

The network model and the related new mode of knowledge production for applied scientific research challenges planners and planning units to change their roles and responsibilities in order to help their organizations face new challenges. Planners become strategic leaders in their organization's planning network and planning units evolve to function as the strategic, central node of such a network. The most important challenges for planners and planning units in this context are touched upon here.

Decentralizing and maintaining the planning process

Flexible planning implies decentralizing planning to the level of individual nodes in the network. Planners play the role of coordinators and catalysts in the general planning network, as well as the roles of advisor and facilitator to those in the planning network who serve as its internal source of energy. Planning units thus become a strategic, central node in the planning network.

Building and sustaining the planning network

The move to flexible planning requires the creation and maintenance of an organization-wide planning network. Planners lead this process of strategy-making in a variety of roles: as catalysts, advisors, promoters, conceptual thinkers, soft analysts, advocates, negotiators, partnership builders, and network guardians.

Building a planning-network culture

There can be no sustainable planning network without a network-related cultural basis. In many organizations, the traditional planning culture must be reoriented to deal with the requirements of complex networks. Following a change in their attitudes and views on planning, planners should play the role of educators to sensitize staff to the new planning culture.

Networking

Flexible planning demands networking and team-building. Planners become networkers. Their offices are places where they are least likely to be found. They act at the internal, decentralized nodes of the organization's planning network, as well as at the edges of the network where new external nodes may be negotiated and even created.

Capacity building in planning and networking

The performance of planning depends on the planning and networking capacity of participating actors. As trainers, facilitators, and team builders, planners will lead the process of capacity building in planning and networking. As planning in complex decentralized networks of heterogeneous partners is inherently more complex than planning at the level of a single organization, considerable capacity is required.

Brokering knowledge

Planning in networks broadens exchange of knowledge within the organization and, especially, between the organization and other potential research partners. This exchange requires professionals with negotiating and brokering skills. Planners should be among the organization's key knowledge brokers and negotiators. They work at the internal and external network nodes where knowledge exchange has its greatest potential.

Scanning the future

Uncertainty is the premise of flexible planning. Planners lead the process of scanning the future, prospecting for demands that may soon require institutional as well as technological innovations. Planners need to master techniques for scenario-building and carrying out long-term planning and prospective analysis. They assume the role of what Mintzberg calls "soft analysts," that is, those who do strategic thinking and lead strategic studies relevant to organizational innovation.

The network organization is context-oriented, project-based, and team-dependent. But to function as networks, traditional, rational-model organizations will need to change. The network organization comes close to what Mintzberg (1994) calls the "adhocracy" type of organization – designed to carry out project-based, expert work in highly dynamic settings where actors must collaborate in teams, coordinating activities by mutual adjustment. The emerging roles and responsibilities of planners and planning units depend on the modernization of their organizations. If organizations do not reconfigure themselves to do project-based, context-oriented, transdisciplinary work, then planners will be unable to develop new skills and deliver new contributions. By the same token, without structural and cultural change, planning units will be unable to support the reconfiguration of knowledge, resources, and skills needed to transform existing vertical planning structures into a horizontal planning network capable of fostering organizational innovation.

The transformation of the roles and responsibilities of planners and planning units will be an integral part of strategically managed organizational change.

Planners and planning units may even have to change their job titles, to describe more accurately their increasingly multifunctional roles and responsibilities. As leaders of the organization's planning network, planners will become strategic thinkers, in addition to being managers and researchers. Planning units, as the central node of the planning network, will be turned into strategic "think-tank"-like units.

Examples

Brazil: Strategic management of agricultural research in Embrapa

Embrapa, the Brazilian Corporation for Agricultural Research, is one of the largest agricultural research institutes in the world. At its 41 research centers it employed in 1993 more than 9700 employees, of whom more than 2000 are managers and scientists. Embrapa was created in 1972 in a government effort to redesign the country's development model and the nation's institutional matrix. Embrapa emerged as a product of Brazil's social, economic, political, technological, and institutional realities in the 1970s. It stressed centralized planning and management, developing a research agenda driven mainly by researchers' interests. National priorities prevailed over ecoregional concerns; interaction with the external environment was minimal. A "productivist" paradigm shaped Embrapa's view on the role of agricultural research and technology.

While successful during the 1970s, Embrapa's performance, together with that of other public institutions, was called into question in the 1980s. By 1990, the need for institutional change was felt, to realign Embrapa within the massive environmental, social, economic, political, scientific, technological, and institutional changes that were taking place in Brazilian society. Following a strategic management approach, Embrapa set out in search of a new paradigm to guide its institutional policies and staff. A stepwise process was followed. Documents signaling the need for change were reviewed. Future scenarios for agricultural research were developed. A secretariat was established to manage the change process. Guiding principles were formulated for the change process and objectives were published and disseminated. Key actors were then trained in the change process. Each research center drafted a strategic plan in collaboration with stakeholders, and those plans were subsequently revamped for consistency. A process then began of international evaluation, political negotiation and adaptation of the overall plan, and finally, implementation.

In the process, Embrapa defined its new institutional paradigm as "enterprise with social accountability." This new paradigm stresses social, economic, and environmental commitment, the values of total quality management, a demand-oriented approach to defining the research agenda for agriculture and agroindustry; holistic perspectives on approaching problems and finding solutions; and administrative transparency and accountability.

The Netherlands: the National Council for Agricultural Research

The Netherlands reorganized its National Council for Agricultural Research (NRLO) in 1995. NRLO "new style" was intended to serve two purposes: to reinforce a long-term perspective on agricultural research policy and to increase innovation in agricultural research. NRLO's mission is to develop a long view on advances that stimulate sustainable development of agribusiness, rural areas, and the fishing industry. It does this through strategic foresight studies. Before its reorganization, NRLO's main task was straightforward and planning-oriented. It was to contribute to the agricultural research system's mid-term programming cycle with a time horizon of five to 10 years. Before the reorganization, NRLO followed a traditional public-sector approach, working through a large number of (fixed) expert committees. After 1995, it began to rely more heavily on temporary, flexible networks, including a broader range of stakeholders.

In its foresight studies, the council has moved beyond traditional, linear models of technical change and innovation, whether science- or demand-based. NRLO recently developed a new, interactive paradigm based on the idea that there are three separate but interdependent domains of knowledge creation: production of (fundamental) knowledge, development of technology, and innovation.

NRLO's "new paradigm" is that realizing sustainable development in the Dutch agricultural sector will require profound and complex innovations. These are referred to as "system innovations":

- designing and introducing entirely new systems rather than merely improving existing ones, using an approach that transcends disciplinary boundaries
- demanding new innovation-creating networks, uniting heterogeneous parties – from both within and outside agribusiness – in concerted action
- encouraging researchers, government agencies, and the business community to become more dynamic in their operations, breaking away from their traditional modes of behavior

References

Castells, M. 1996. The Rise of the Network Society. The Information Age: Economy, Society and Culture. Vol. 1. Oxford: Blackwell Publishers.

Castells, M. 1997. The Power of Identity. The Information Age: Economy, Society and Culture. Vol. 2. Oxford: Blackwell Publishers.

Castells, Manuel. 1998. End of Millennium. The Information Age: Economy, Society and Culture. Vol. 3. Oxford: Blackwell Publishers.

Chisholm, R.F. 1996. On the meaning of networks. *Group & Organization Management*, 21 (2): 216–235.

Gibbons, M., et al. 1994. The New Production of Knowledge: The Dynamics of Science and Research in Contemporary Societies. London: SAGE Publications.
Khun, T. 1970. The Structure of Scientific Revolutions. Chicago: The University of Chicago Press.
Mintzberg, H. 1994. The Rise and Fall of Strategic Planning. New York: Prentice Hall.
Tabor, S. R. 2000. Globalization: Planning agricultural research in an open market economy. In *Sourcebook on Planning in Agricultural Research.* The Hague: International Service for National Agricultural Research.

Chapter 16
Participation in Agricultural Research Planning

Louise Sperling and Jacqueline Ashby

Participation in agricultural research planning can help achieve a range of objectives: greater relevancy of research, representativeness, equity, refined insights, and broadened ownership of the research process. It can also contribute to the democratization of research, particularly for the poor and marginalized. Types of participation vary from "passive" to "decision-making" according to the objectives achievable. Factors directly affecting the quality of results include choice of the type of stakeholders to involve, decisions as to who will represent them, the participation process itself, and the overall strategy for stakeholder involvement (centralized, decentralized, or by contract).

What is participation in agricultural research planning?

Participation in agricultural research planning implies that stakeholders are involved in setting research agendas. Stakeholders may be active at different levels of planning (e.g., national, regional, or local) and at different stages (e.g., setting broad agricultural sector goals, developing agroecological strategies, or defining community research priorities). The stakeholders emphasized in this chapter are those near the end of the research planning chain, that is, farmers, consumers, and traders. Ultimately, the effectiveness of agricultural research planning is defined by whether such "end users" do benefit from research. End user involvement in research planning implies many challenges: such end users tend to be heterogenous, not organized into formal groupings (which might speak for them), and are often culturally and economically distinct from the dominant management groups in agricultural research organizations.

Objectives of participation

There are compelling reasons for involving stakeholders, especially end users, in agricultural research planning. These range from technical imperatives (gaining insights) and equity concerns (reaching needy constituencies) to ensuring collaboration in subsequent research phases. Objectives of user involvement can be sketched as follows:

- *relevancy* – aiming to bring about more demand-driven and client-oriented research and extension, perhaps leading to more effective use of research resources.
- *representativeness* – to enable research programs to reach broad and varied constituencies. Direct involvement of various groups often leads to articulation of distinct wants and needs. One research direction does not fit all. In practice, addressing issues of representativeness leads to greater coordination among and within stakeholder groups.
- *equity* – to address concerns of the more marginalized stakeholders. Disadvantaged sectors of the population are seldom well served by agricultural research organizations. Only recognition of their needs in the planning phase will increase the possibility that their distinct concerns will be addressed.
- *research insights* – to gain from the technical and social insights of those close to specialized research issues. Stakeholders may add precision in defining researchable constraints and their assessments of what is feasible may improve the quality of research projects, as well as reduce the number of dead-end projects.
- *ownership* – to bring on board the range of stakeholders needed for the success of a technical innovation. Research proceeds more efficiently and effectively if those implicated have a voice in planning it. Ownership of the research process enables work at much larger scales and with greater geographic coverage and can increase the potential for longer term projects. True ownership also implies some form of cost sharing in the research process.
- *democracy* – to respond to the wave of democratization and decentralization. This rationale focuses on the empowerment of people, rather than on research products per se.

Types of participation

Stakeholders can participate in agricultural research planning to different degrees and in different roles. Participation may range from simple surveys of or meetings with farmers' groups on their wants and needs, to collaborative forms of planning, through to planning efforts in which end users have real decision-making power to select priorities. Participation in selecting priorities is often linked to forms of mutual commitment such as joint evaluation and accountability sharing.

Using the example of one type of stakeholder, farmers, table 1 suggests different ways in which participation might take place in agricultural research planning. While consultative types of participation are becoming common, more profound types of joint decision making are still rare. Some of the more

Table 1. Types and Features of Farmer Participation in Agricultural Research Planning

Type of Participation	**Key Features**
Passive participation	Farmers listen to what is going to happen or has already happened. This is a one-way announcement by an administration or management unit. No response is elicited.
Participation by giving information	Farmers answer questions posed by researchers. Interpretation and synthesis is left to those raising the issues. Farmers do not necessarily have the opportunity to influence the shape of the proceedings.
Participation by consultation	Farmers are consulted and external people listen to their views. These external professionals define the problems and solutions, both of which may be modified in light of people's responses. Such a process does not automatically concede a share in decision making; yet it often leaves professionals better informed. Professionals are under no obligation to accept farmers' views.
Collaborative or interactive participation	Farmers participate in joint analysis which leads to agreed action plans. There is an exchanging and synthesizing of ideas articulated by the different groups involved.
Decision-making participation	Farmers have final say over certain aspects of the research agenda. They also take greater accountability for following through on action plans.

Source: Adapted from Pretty and Chambers 1994.

empowering participative planning approaches are now being used to plan agricultural research addressing complex concerns, such as integrated land and water management and pest management. Means of participation are also being explored on scales that were previously difficult or seemingly intractable for agricultural research to operationalize, for example, at the watershed, steep-hillside, and river-basin levels. This evolution stems from the recognition of planning as part of a cycle in which stakeholder interaction is key, not optional, for achieving positive agricultural research impacts. Hereby planning itself is seen to need modification or even redirection in response to what occurs in the implementation phase of a research project. The success of this process of ongoing evaluation and adaptation largely depends on the quality of stakeholder involvement.

The type of participation to strive for depends on the objectives of the planning process. If "improving technical insights" is considered the main objective (and questions of ownership or equity are not put at the forefront), a consultative form of participation in planning might be deemed adequate. *Consultative participation* seldom leads to agreements on subsequent actions. It tends to be relatively rapid, and is not necessarily linked to a feedback phase in which research plans are adapted. However, one advantage is that consultative types of participation are relatively easy to program and achieve. If objectives such as equity or ownership are aims in agricultural research planning, more encompassing forms of participation may be called for. Equally, if agricultural planning itself is conceived as a cyclical process that links planning to subsequent action and evaluation, more encompassing forms such as "collaborative participation" should be sought. *Collaborative participation* takes more time and skill. It also demands that a process be in place to mediate among different stakeholders and, sometimes, resolve conflicting views or priorities. *Decision-making participation* can be implemented only if the resources available to the different stakeholders in the research priority-setting process have been pre-established. This implies a clear policy for determining which stakeholders will be represented in the planning process, the relative weight of their respective needs and wants, and the resources for which they ultimately will be responsible. This is the most challenging type of participation to achieve, but the one which has the potential to achieve all the objectives set out in the previous section.

Enhancing participation in agricultural planning

Several key decisions shape the participation process in agricultural research planning. Choosing whom to involve and how to involve them are among the more basic concerns.

Deciding who: which types of stakeholders

There are many different types of stakeholders who should be involved in agricultural research planning. Some have a more direct stake than others, as can be shown in the case of new variety development: immediate users include farmers, traders, consumers, and plant breeders. Less direct users, but who are still very tied to the technology, are varietal release committees, seed services, and extensionists. Still less direct (or indirect) stakeholders are those who may be affected by the spin-offs of use of the technology. For plant varieties, these may include environmentalists concerned about biodiversity and preserving local germplasm.

Three questions should guide the choice of which types of stakeholders to consider involving: "Whom do you want to benefit?" "Who has specialist insights which might help shape research directions?" And, "Who will be affected by the spread of research results and technologies?"

Table 2 suggests groups of stakeholders to include in research planning. Each country and context, however, will have to tailor its own definition of beneficiaries and stakeholders. Much depends on the type of agricultural research being planned (e.g., cash crop production, natural resource management research, or inquiry into food security issues).

Table 2. Categories and Groups of Stakeholders to Include in Research Planning

General Category of Stakeholders	Examples of Specific Groups
Direct beneficiaries/stakeholders	– farmers: low income, middle income, and wealthy – pastoralists and foragers – intermediaries, such as processors and traders – consumers (Note: representation has to include the different agroecological zones and socioeconomic groups, such as wealth, gender, ethnic group)
Stakeholders with specialist knowledge	– actors from the research and extension system – farmers' group representatives (farmers' research organizations, women's groups, specialist producer cooperatives) – agricultural policy analysts (government and private sector) – economic policy analysts (government and private sector) (Note: for research planning, adequate representation of regional differences is important)
Indirect beneficiaries/stakeholders	– environmentalists representing urban public – nonagricultural rural groups

In thinking about stakeholders in planning, it is important to factor in interests of future as well as present generations. Some stakeholders, such as smallholder farmers, often have short-term research and development priorities, which, taken alone, may divert research from long-term, less immediately tangible benefits such as environmental conservation. Further, even in low-income countries, social interest groups other than smallholders may have valid current needs (which have long-term consequences), for example, consumers may be concerned about pesticide residues.

Identifying representatives for stakeholder groups

Never can all members of each stakeholder group be involved in agricultural planning, no matter what the scale of the planning or the location. There may be established procedures for selecting representatives from formally organized groups, such as government ministries or extension services. For instance, the government may have a specially designed policy analyst whose field is agricultural research. The challenge with organized groups may be to pinpoint representatives who have both the political clout to convey a viewpoint and specialist knowledge of key agricultural concerns.

Many stakeholder groups, however, and particularly farmers, may be unorganized and without identifiable representatives. Moreover, they may be internally quite heterogeneous. There are documented cases where upper-class or export-oriented farmers have been able to influence research budgets and effectively lobby for specific technologies to be developed. But poor farmers, particularly those who are less market-oriented, organize less easily, and while their real ability to say "no" to a technological agenda is felt, it may be felt erratically (Röling 1989).

As a result of this "gap," there is often a role for an intermediary institution in planning to translate local voices for policymakers and those managing the priority-setting process. Community-based organizations, farmers' organizations, and even intermediary nongovernmental organizations may be asked to fill this liaison role. While such agencies have recognized limitations in speaking for farmers, under optimal conditions they do carry the voice of beneficiaries through a filtering and structuring process that makes it intelligible to other audiences (see Holland and Blackburn 1998).

Strategies to get stakeholders involved

Having understood who the stakeholders are and identified appropriate representatives, the decision remains about how to devise a valid process of planning. Several different mechanisms, at different levels of agricultural research, have given initial, effective results. All have weaknesses as well as strengths that need to be addressed. Much depends on the scale of planning and the diversity of the stakeholder group.

Centralized planning with indirect "proxy" representation. The planning strategy that most parallels what a research system routinely does is to bring an array of stakeholders into the arena where centralized research planning takes place. Thus representatives from groups like ministries, research and extension, nongovernmental organizations, and farmers' organizations might meet to plan and debate a specific research agenda item. The challenge is to infuse such debate with a set of tools that are meaningful for all. But the process is often too abstract for those most profoundly affected, farmers, to become fully engaged. Intermediaries often serve as proxies for local-level groups, with expression of farmers' true viewpoints dependent on the intermediaries' abilities to aggregate needs and wants of diverse farmer constituencies and comprehend local priorities.

Decentralized planning with direct representation by local stakeholders. Decentralized planning offers a second potential option for participation in planning with a wide range of groups. Herein a series of local-level meetings are organized, followed by a synthesis and funneling of information gathered to an administrative center (see the Tanzania case in the examples below). Participatory tools can be very effective for highlighting community planning preferences (assuming power biases can be mitigated). And various approaches are available for overcoming the potential challenges of use of diverse languages (including western technical and conceptual languages). While the local-level insights gained here can be quite refined, the hitch often comes when trying to synthesize and sufficiently homogenize planning outcomes from many sites. Information gained may have site-specific validity, but be considered too "anecdotal" for steering research priorities at higher regional or national levels (Tripp 1991).

Both the above planning scenarios can generate critical information about stakeholders' key problems and fine-tune a research system's ability to plan and implement research. Yet planning with a wide stakeholder group *cannot guarantee* that the people who make decisions about the content of research will actually use this knowledge and information (Merrill-Sands and Collion 1994). Under both scenarios, the issue of taking different stakeholders seriously hinges, to a large degree, on researchers' good will, with a substantial dose of interpretation as to stakeholders' real wants and needs (Ashby and Sperling 1995).

Contractual research planning. A different mechanism for determining whose research priorities are given weight in agricultural planning is one that places a significant proportion of the resources available for financing research directly under stakeholder control. This approach removes the need for direct face-to-face multi-stakeholder planning by creating the means for stakeholder groups to contract applied research and so exert a demand pull on the research system. This approach has worked quite well in the North, in countries such as the Netherlands. There organized groups of farmers participate in government program committees, have representatives on boards of main agricultural organizations (Röling 1990) and directly pay almost half the costs of experiment stations and farms. It has also been successful in the South among the more wealthy, export-oriented farmers (e.g., cotton growers in Côte d'Ivoire and sugar farmers in South Africa) (Carney 1998) and in plantation crops for which the number of growers is small but well informed. The approach is relatively novel, though, in use with resource-poor farmers. In Mali, participation of poor farmers in agricultural research priority setting is heavily subsidized by northern donors (Collion and Rondot 1998).

In planning contractual research, a "reverse" participation problem may arise. If farmers' groups fail to consult with researchers when defining their agendas, a research strategy may emerge that has limited potential for technological progress, precisely because the subjects selected may not be amenable to technical investigation. In addition, the research agenda may to be too "farmer driven" and lack the political support of other stakeholders, including consumers and environmental groups. In this situation, one bias is simply exchanged for another, still without the ability to define beforehand whether the exercise of contracting improves research performance and efficiency.

The above strategies for involving stakeholders in research planning represent a continuum from more voluntaristic types of consultation (and more informal, nonbinding commitments) to strict governance by stakeholders of the planning process – and results – through financial leverage. In the latter case stakeholders are most likely to get the product(s) they desire.

Relevance for agricultural research

Participation in planning is key for ensuring accountability and relevance in agricultural research. There are, however, a number of prerequisites for enhanced stakeholder involvement which should be anticipated in the research planning process.

Clarify expectations

From the very beginning, all stakeholders need to know what is expected of them and what they can expect from the process. Stakeholders should be aware in advance of what information the group needs from them, procedures for meetings, how input will be subsequently used, and what follow-up actions to expect. For example, "Will giving information lead to decision making?" "Are resources being offered or guaranteed?" And, "Is this a one-off consultation or the beginning of an ongoing collaborative process?" These elements need to be negotiated and agreed on *beforehand* and monitored in the planning process.

Prepare capacity to respond to diverse agendas

Stakeholder-driven agendas are likely to differ markedly from those geared toward basic, long-term research. Addressing stakeholder needs means that the agricultural research planning process itself must be sufficiently decentralized and flexible to meet stakeholder goals and to encourage site-specific, local planning.

Address competing interests and resolve conflict

The more stakeholder groups are involved, the greater the likelihood that a big shopping list of demands will be handed over to the research system. The planning process, however, can anticipate this. An institutional will and ability should be cultivated to resolve differences and compromise. Use of a facilitator (see below) is helpful in this realm.

Build in accountability

Mechanisms of accountability have to be built into the planning process, to ensure that planning and its subsequent actions did indeed meet stakeholders' needs. Accountability can be built in multiple ways. Those involved in the original planning sessions may be invited on a predetermined basis to give feedback on the unfolding research agenda. Or those involved in planning could have channels through which to register concerns and suggestions. Groups involved in the planning process should have the option *not* to be involved in the future should their needs not be met.

Factor in sufficient time

Different stakeholders need time to develop their own group positions and to assimilate the wants and needs of other groups, who may hold substantially different assessments of what research programs should have priority. Time should be planned into the process for interaction with home constituencies.

Factor in sufficient resources

Bringing people together at one site or holding many meetings in a number of locations requires support funds. One could well argue that such planning saves money in the long run through more targeted research and more efficient divisions of labor. But the fact remains that start-up funds are needed to fuel the process.

Use flexible planning methods

While researchers and policymakers may be familiar with similar sets of planning tools, farming communities definitely are not. Some tools that have proved useful in facilitating participation are those that visualize discussions, such as cards or diagrams. Visualization demands that each group be clear on what they want to say and allows relationships among themes to be mapped. For illiterate groups, symbols may be useful for visualization. But even for literate groups, simple, mutually understandable language should be used, or even several languages simultaneously if necessary.

Arrange for a neutral facilitator

Professional facilitators are good to have on hand during planning sessions. These individuals should be adept in diverse communication tools, able to counter dominance by certain groups, and skilled in verifying (using sophisticated cross-checking) that diverse groups interpret the agreed conclusions in the same way.

Anticipate logistical needs

Meetings should be planned when key stakeholders can truly focus on the challenge at hand (the peak weeks of harvest would not be optimal for farmers or field researchers). Language barriers should be anticipated and mitigated. Cultural and socioeconomic sensitivities should be respected (e.g., if women can't meet together with men in public, several sets of meetings should be undertaken).

Examples

Morocco: centralized decision making, centralized planning

Morocco's centralized research planning approach has proven useful, for instance, for a commodity program (e.g., rice), an agroecological zone (e.g., a highlands program), or a production system (e.g., intensive animal husbandry). It uses two workshops: the first lasting three days, the second two days. Participants are experienced scientists from the national agricultural research organization, international agricultural research institutes, universities, and government ministries. Successful use of this approach in Morocco's barley research program is described in Collion and Kissi 1995 (see also Collion, this volume).

Tanzania: Decentralized planning using regional teams

In Tanzania's farming systems research program, researchers facilitated zone-by-zone workshops (a total of seven) involving local scientists, extensionists, nongovernmental organization representatives, and a small number of farmers. The approach drew on the German "ZOPP" method, or planning by objectives, in which a set of problems is first identified, then translated into researchable objectives, and a subsequent action plan developed. Planning results at each location are then compared and made compatible at the central level.

A similar approach is being implemented on a more permanent basis in a newly developed African highland initiative, which involves stakeholder groups in planning as well as in more durable steering committees. These committees are fora for regional coordination and information sharing. They also allow discussion of in-depth planning and technical follow-up.

Côte d'Ivoire: contracted models

The Compagnie Ivoirienne pour le Développment des Textiles (CIDT), the Ivorian cotton development agency, has signed a financial agreement with the industrial crops department of the Institut des Savannes (IDESSA), a regional research institute. Through the agreement, IDESSA provides technological backup to CIDT. The two institutions collaborate closely in research, from planning to technology release. Planning meetings, technical committees, joint trials and field visits, and liaison positions, all defined under the agreement are used as linkage mechanisms. Owing to this collaboration, cotton is the subsector that has experienced the most success as far as small producers are concerned. Cotton yield has more than tripled over the last 30 years in the savannah zones of Côte d'Ivoire (Eponou 1993).

References

Ashby, J. A. and L. Sperling. 1995. Institutionalising participatory, client-driven research and technology development in agriculture. *Development and Change*, 26 (4): 753–770.

Carney, D. 1998. Changing Public and Private Roles in Agricultural Service Provision. London: Overseas Development Institute.

Collion, M. -H. and A. Kissi. 1995. Guide to Program Planning and Priority Setting. The Hague: International Service for National Agricultural Research.

Collion, M. -H. and P. Rondot. 1998. Partnerships between Agricultural Services Institutions and Producers' Organisations: Myth or Reality. Agricultural Research and Extension Network, Paper No. 80. London: Overseas Development Institute.

Eponou, T. 1993. Partners in Agricultural Technology: Linking Research and Technology Transfer to Serve Farmers. Research Report No. 1. The Hague: International Service for National Agricultural Research.

Holland, J. and J. Blackburn (eds.). 1998. Whose Voice? Participatory Research and Policy Change. London: Intermediate Technology Publications.

Merrill-Sands, D. and M. –H. Collion. 1994. Farmers and researchers: The road to partnership. *Agriculture and Human Values,* 11 (2 and 3): 26–37.

Pretty, J. N. and R. Chambers. 1994. Towards a learning paradigm: New professionalism and institutions for agriculture. In *Beyond Farmer First: Rural People's Knowledge, Agricultural Research and Extension Practice* edited by I. Scoones and J. Thompson, pp. 182–203. London: Intermediate Technology Publications.

Röling, N. 1989. Why farmers matter: The role of user participation in technology development and delivery. In *Making the Link between Agricultural Research and Technology Users*. Boulder, Colorado: Westview Press.

Röling, N. 1990. The agricultural research-technology transfer interface: A knowledge systems perspective. In *Making the Link between Agricultural Research and Technology Transfer in Developing Countries* edited by D. Kaimowitz. Boulder, Colorado: Westview Press.

Tripp, R. 1991. The farming systems research movement and on-farm research. In *Planned Change in Farming Systems: Progress in On-farm Research* edited by T. Tripp. Chichester: John Wiley & Sons.

Website

http://www.ids.susx.ac.uk/ids/partip/intro/introind.html

Chapter 17
Priority Setting

Rudolf B. Contant

Priority setting is the process of defining a research portfolio that is consistent with the country's agricultural policy, the research organization's mission, and the research program's objectives. Priorities are normally set in the light of limited resources, increasingly diverse research needs, and growing demands for transparency in resource allocation. This chapter discusses priority setting in the context of planning. Tools for systematic priority setting are briefly described and indicators identified to assess the quality of a priority-setting exercise .

What is priority setting?

Agricultural research planning must respond to the stated needs of and opportunities in the agricultural sector and its subsectors. Priority setting is the final phase of the research planning process. It enables planners to define a research portfolio that is consistent with the country's agricultural policy, the research organization's mission, and the research program's objectives. The context in which priority setting is done is characterized by limited resources, increasingly diverse research needs, external demands for greater transparency in resource allocation, and strengthening focus on client needs.

A research priority-setting exercise takes its cues from a sector or subsector plan. This plan describes the main features of the (sub)sector, reviews the main policies and macroeconomic reforms that influence it, and summarizes the government's (sub)sector strategy. The objectives of such a plan are often phrased in terms of food security, poverty alleviation, income generation, and natural resource sustainability. The overall national agricultural research plan is designed to contribute to all of these development objectives, while a program plan within it may contribute to one or several of them.

Before any research priority-setting exercise can get underway, clear answers are needed regarding the level at which priorities are to be set, who should be involved in setting the priorities, and how much time and money is available for the exercise.

At what level are priorities set?

Three levels are usually distinguished in research planning and priority setting: the national level, the level of an individual research organization, and the program level. At the national level, the relative importance of major research areas is determined. At the level of a research organization, decisions are made on the place of different research programs in its portfolio. In program-level exercises, decisions are made on the relative importance of possible themes within a research program.

Who should be involved in setting priorities?

The quality of planning as a whole and priority setting in particular depends to a large extent on the participation of representatives of principal clients and stakeholders. This group must be large enough to represent main interests and small enough to permit effective communication. At the program level, strong input is required from farmers' and grassroots organizations (the process should incorporate ways for less articulate stakeholders to participate actively). At the research organization level, senior research managers and stakeholder representatives should play a prominent role. At the national level, politicians and policymakers have a central role to play. Other groups involved are food processors, market brokers, input suppliers, extension agents, nongovernmental organizations, and consumer organizations.

How much time and money may be spent on the exercise?

Priority setting should be cost effective, but an organization should not fear investing some time and money in it. Practices vary. In some places, a great deal of time is invested in planning, resulting in less flexibility in implementation. Elsewhere decision making is quicker, with the understanding that plans can be modified along the way. Whatever the practice, there should be a reasonable balance between resources committed to priority setting (and planning), and those available for project and program implementation. The balance also depends on the increase in research benefits that is expected as a result of the priority-setting effort.

It appears reasonable to expect that by putting 5% of available resources in planning, the remaining 95% of resources will be used to more effect than if 100% were spent on research operations alone with no planning applied. If 5% for planning is taken as is a reasonable estimate, perhaps priority setting should take one-third of that (i.e., under 2%). The level of investment to be made in compiling and, sometimes, collecting information for priority setting has to be determined early on. Otherwise disappointments are likely if information proves unavailable or too expensive to obtain. Information can come from published data and reports, as well as from scientists and extension workers, research clients, and other key informants.

Steps in priority setting

Priority-setting exercises should incorporate the following steps:
- identifying the range of research alternatives among which to set priorities
- defining and applying the priority-setting method with its criteria and measurement indicators
- sensitivity analysis
- linking priorities to resource allocation
- validation and preparation for implementation

Identifying the range of alternatives

No priorities can be set without having first decided on the alternatives among which choices are to be made. Although this seems obvious, it is the most neglected aspect of priority-setting exercises. Often, research alternatives are taken to be simply a list of possible subjects that comes to mind (e.g., all the country's crops and a few major thematic areas such as "soils" or "drought tolerance"). This is hardly satisfactory. It is preferable to derive research alternatives from an analysis of constraints and opportunities, for the country as a whole or by zones, based on the needs and opportunities of specified user groups. Within programs, research projects need not be elaborated in detail, but their overall operational costs, time requirements, anticipated outcomes, and benefits should be determined. Only then does it become possible to make predictions on the relative impact of research and to compare and prioritize alternative lines of inquiry.

Selecting a method, criteria, and indicators

A wide range of priority-setting methodologies is available, of varying complexity. It is an art to choose the right method for the circumstances and the decision problem at hand. Priority-setting methods differ considerably with regard to data and analytical skills required and with regard to the participation of decision makers in the exercise. Quantitative, analytical methods often rely on outside technical analysts, without direct involvement of key decision makers. This may create a problem of credibility, as decision makers do not understand the methodology and have little confidence in the results. More participatory approaches, however, often lack rigor and stability, that is, they cannot be replicated with similar results.

To prioritize research alternatives, criteria are needed to evaluate the potential contribution of a particular activity to the research objectives identified earlier in the planning process. If several criteria are used, they may receive different weights. Good criteria are of practical use only if they can be measured with enough precision to make transparent discrimination possible between research alternatives.

Sensitivity analysis

The initial outcome of priority setting is often submitted to a test of its sensitivity to changed assumptions. This can be done through group analysis and discussion or by mathematical procedures in which the measurement method, the criteria weights, or parameter values (prices, unit cost, yield gain, probability of research success, probability of adoption) are modified. This may lead to reasoned changes in the order of the priority ranking.

Linking priorities to resource allocation

The relationship between setting priorities and allocating resources is not always straightforward. Three issues tend to complicate the process: uncertainty about resource requirements, resources that are "locked" in ongoing programs and projects, and the influence of interest groups.

Priorities in relation to resource requirements. Priorities cannot always be translated directly into an allocation of resources. Here two different situations may arise. The first is when resource requirements have not yet been defined, which is often the case in setting national-level priorities among commodities, programs, or regions. In this situation, resource allocation should follow the priority ranking. If research program A has higher priority than program B, and program B has higher priority than program C, the amount of resources for program A should be higher than for B, and the allocation for B higher than for C. If this is currently not the case, research managers can easily see which programs might require more resources, which should be left stable, and which should be reduced in size.

The second situation is when resource requirements have already been defined. This is often the case when setting priorities among projects where expected resource needs are known. For example, if project ***a*** has priority over ***b***, and ***b*** over ***c***, normally project ***a*** will be implemented first, followed by project ***b*** and ***c*** if sufficient funding is available. Several factors may complicate the application of this decision rule. Lumpiness in project requirements may cause a low priority project with small resource requirements to be implemented before a high-priority project that requires more resources than are available. Lack of resources other than funding, for example, specialized staff or equipment, may lead a lower priority project to be favored temporarily, until the resource constraint for the higher priority project is overcome. Highlighting gaps in the resources available to the organization is an important result of priority setting, and may influence resource development and acquisition strategies.

Resources are "locked" in ongoing programs and projects. A change in research priorities may not be implemented immediately because the resources invested in ongoing research cannot be feasibly redirected towards another purpose, or because it takes time to assemble the resources required. For example, animal breeding programs require a long-term perspective, both in terms of scientific disciplines and animal stock. Discontinuing or scaling up animal breeding programs in response to changes in priorities may therefore take some time.

Influence of interest groups. Resource allocation is a political process that often entails conflict. Agreed research priorities may not be implemented immediately because they do not reflect the objectives of some powerful interest group. Certain (commercial) farmers may be better organized than other (smallholder) farmers and may exercise more influence over the research agenda. Since many research organizations have limited operational funds, their research agenda may be swayed by a small amount of operational funding provided by a financier with specific objectives or priorities. Such deviations from agreed priorities may be held in check if the research portfolio has been developed in a broadly based and robust process, and if there is consensus on the need to implement the agreed portfolio.

Validation and preparation for implementation

Results obtained in the priority-setting exercise should be subjected to wider validation at various stages during the planning and priority-setting process, for example, through meetings with a larger group of stakeholders. Stakeholder validation improves the quality and relevance of the results, but also helps to build consensus on the research portfolio and to facilitate implementation of the selected priority activities. Validation sets the stage for development of the detailed implementation plan.

Doing priority setting

This section examines some specific aspects of the research priority-setting process.

Team building and deciding on participation

The first step is to build a small priority-setting core team of persons able and willing to dedicate to it a significant share of their time. This team should be closely linked to, or made up of, the people responsible for the overall planning exercise. It organizes meetings, prepares background information, and assumes responsibility for computing the results of the exercise and sharing results with the wider group of participants. The team leader (e.g., the program director) ensures that priorities are established in accordance with national objectives and is in active charge of the process as a whole. The team should have at least two other members: a socioeconomist and a technical scientist. A facilitator may also be required to manage meetings.

The degree and extent of participation of different stakeholders varies according to the level at which priority setting takes place. For some client groups, it is a challenge to identify persons who can validly represent their group's needs. Representatives of end users, researchers, and decision makers should participate, bringing together a wide range of technical expertise. User representatives may be key farmers or leaders of farmers' organizations. Extension services and nongovernmental organizations may represent end users if they understand their problems well; they have the additional advantage of being able to be partners in research as well. Experienced researchers can provide information on the technical feasibility of research alternatives. They do not necessarily belong to the organization for which the priorities are being set. Decision makers (top managers, board members, and senior government officials) are essential participants, because they are ultimately responsible for the priorities arrived at. Although they cannot be deeply involved in the exercise, their role is particularly important at two stages: (i) when the objectives of research and their relative importance (weights) are being determined and (ii) when validating the final outcomes.

A word of warning is in order with regard to the common practice of having priorities set by the planning unit, which then forwards its recommendations to the (research) director. Priorities set without the explicit support of research program staff and project leaders are likely to lead to conflict within the organization.

Defining research alternatives

Priority setting is best based on a formal analysis of constraints and opportunities. Constraints analysis and priority setting may be done by target zone, in which case the zones are derived by spatial analysis using a geographic information system (GIS). An essential ingredient is knowing the needs of user groups (identified, for instance, by rural appraisals and consultation meetings). Constraints and needs analysis lead to project identification, a first characterization of the research alternatives from which priorities will be later selected.

Choosing a priority-setting method

A wide range of methods and tools for priority setting are available. Among the single-criterion tools are congruence, benefit/cost analysis, and economic surplus analysis. The principal multiple-criteria tools are simple checklists, various forms of scoring (with or without weights attached to different objectives), analytic hierarchy process, and mathematical programming. Single-criterion priority-setting tools may be incorporated into multiple-criteria tools. There is abundant literature describing the strengths and weaknesses of each of these approaches, and some references are recommended at the end of this section. Table 1 summarizes some of the more popular priority-setting methods.

Institutionalizing priority setting

All agricultural research organizations need some capacity for priority setting. After an approach suitable to the organization is chosen, formal staff training may be required. Capacity is most effectively developed if training is followed by immediate application. Investments in information may also be needed. A socioeconomics or planning unit is often a suitable home for developing and maintaining a capacity for leading priority-setting exercises.

Relevance for agricultural research

A planning process is incomplete without priority setting. Conversely, priority setting makes no sense if not applied in the context of planning. Mandate, scope, and approximate resource availability for the institute or program must be defined, otherwise few people will be committed to the outcomes of priority-setting work.

When developing short-term research plans, priority setting is a useful way to choose between projects that can be readily implemented. However, priority setting falls onto the most fertile ground when the planning mandate is more

Table 1. Summary of Priority-Setting Methods

	Transparency	Participation	Simplicity	Theoretical Logic	Discriminating Potential	Cheap to Apply
A: Single-criterion methods						
– Congruency	***	*	***	*	*	***
– Economic surplus	*	*	*	**	***	*
B: Multiple-criteria methods						
– Multiple models	***	***	**	**	**	**
– Objective programming	*	*	*	**	**	*

Key: * = poor
** = intermediate
*** = good

Source: Janssen 1995.

strategic or long term. Predicting research impact and reviewing how emphasis on different sector objectives would affect the research agenda may clarify the mission of the organization and help it to define a new strategy. If the organization recognizes external pressures in its plans and priorities, this will improve the relevance of its strategy.

The usefulness and quality of a priority-setting exercise can be judged by different indicators, such as stakeholder involvement, choice of criteria, and weighting procedures used.

Involvement of stakeholders

Stakeholders are a varied group, including farmers, farmers' organizations, input and output agents, market brokers, consumers, extension agents, agricultural policymakers, and politicians. Their explicit endorsement of the outcome of the priority-setting exercise is perhaps the most valuable guarantee of the relevance of the emerging research agenda.

Choice of priority-setting criteria

Good criteria have two attributes: (i) they are logically related to stated objectives of the research organization and (ii) they are supported by indicators that credibly discriminate the impact of research alternatives. The most commonly used criteria are efficiency, equity, sustainability, and food security. However, they need to be applied with understanding. Efficiency criteria are always important, as no organization or system can afford to neglect efficiency. Equity criteria focus on welfare gains for certain target groups, for example, the poorest category of farmers. Equity concerns can be incorporated by the choice of commodity or region targeted by the research program. Sustainability criteria examine the contribution of research to objectives related to the maintenance of the natural-resource base for future use. Embedded in sustainability criteria is the question of how society should trade off present benefits for potential future benefits, often under highly uncertain future conditions.

Food security is a criterion to which policymakers attach great importance, but which is often confused with food self-sufficiency. Self-sufficiency objectives lead to agricultural research that favors food crops over cash crops. In doing so, opportunities for trade and export may be lost. Food security objectives aim to lower variability in food supply. To achieve this both domestic food crop production and imports are used in measurements.

Weighting procedures

If a research organization or program wishes to contribute to several sector objectives at the same time, it needs to decide on the relative importance of each.

The extent to which a weighting procedure is applied consistently and satisfactorily is an indicator of the quality of the priority-setting exercise.

Examples

Kenya: priority setting at KARI

Priorities must be set at several levels within an agricultural research organization. For example, the Kenya Agricultural Research Institute (KARI) is structured in three levels: institute (encompassing research stations and programs), program (national and regional), and project. Initially, research priorities at KARI were set by senior management. In 1991, however, institute-wide priorities were reexamined using a scoring model.

In 1994, KARI and ISNAR started developing a process for program-level priority setting that included systematic collection of information that would provide a basis for identifying the potential benefits of specific research themes for targeted zones within Kenya. As part of this process, a priority-setting working group developed initial estimates of research benefits. A program advisory committee, composed of major research program stakeholders, then reviewed the initial results, establishing program priorities. The committee then used these priorities as the basis for its annual review of the relevance of proposed and ongoing projects. Similar priority-setting processes are now being piloted with regional and production factor research programs. The results of these program-level priority-setting exercises will form the basis for a future reevaluation of cross-program priorities at the institute level.

KARI's priority-setting process combines information on client constraints with expert opinions on the potential generation and adoption of technologies and quantitative georeferenced data on climate, soils, population, prices, and production levels. Five steps are followed in synthesizing and using this information in decisions on resource allocation:

1. compiling the information base
2. identifying program research target zones and research themes
3. eliciting the potential for technology generation and adoption
4. ex ante estimation of research-induced benefits
5. establishing priorities with program stakeholders

For further details on Kenya's priority-setting method, see Kamau et al. 1997 and Mills 1998.

Croatia: developing priorities for an agricultural research council

In 1997 the Croatian ministry of agriculture and forestry established the Agricultural Research Council (ARC). ARC manages a competitive grant fund for applied, farm-based agricultural research. The council has 18 members, most of whom are farmers. Research proposals can be submitted to ARC by all research organizations in Croatia, including universities and research organizations.

To ensure maximum impact of its grant scheme, ARC felt that it had to establish priorities at two levels. First, priority subjects for ARC funding had to be defined. ARC preferred a commodity focus over a disciplinary, regional, or production system approach, because it felt a focus on commodities would encourage multidisciplinary, problem-oriented research with a market orientation. Within the commodity focus, ARC selected priority commodities using a scoring model that combined three sets of considerations: economic importance, social contribution, and environmental role of the commodity. Specific, measurable criteria were identified for each consideration, and commodities were assessed according to the criteria. Economic criteria were weighted heavier than the social and environmental criteria. Wheat, beef, milk, poultry, sheep and goat, olive, potato, and bees were selected as initial priority commodities for ARC grants.

With the priority commodities defined, the second level of priority setting was to rank research projects within the priority commodities. A project evaluation system was designed based on three sets of project criteria: importance of the constraint addressed by the project, quality of research, and diffusion potential of results. For each set, specific criteria and ways to measure them were again defined. Farmer representatives evaluate the severity of the constraint; scientists evaluate the quality of research; and assessment of diffusion potential is the responsibility of the extension service. The ARC secretariat manages and oversees the priority-setting process following an annual project cycle.

Recommended reading

Alston, J. M., G. W. Norton, and P. G. Pardey. 1995. Science under Scarcity: Principles and Practice for Agricultural Research Evaluation and Priority Setting. Ithaca, NY: Cornell University Press.

The lengthy volume examines formal economic analysis, economic surplus analysis, econometric techniques, mathematical programming procedures, and scoring models. It explores conceptual foundations of these practices and how to do them.

Braunschweig, T. and W. Janssen. 1998. Establecimiento de prioridades en la investigación biotecnológica mediante el proceso jerárquico analítico: Experiencias en Chile. Research Report No. 14. The Hague: International Service for National Agricultural Research.

This Spanish-language book describes the adaptation and application of the analytic hierarchy process (AHP) for agricultural biotechnology research in Chile. It discusses the advantages and disadvantages of AHP in comparison with other methods and suggests future improvement of this tool (English edition Priority Setting in Biotechnology Research Using the Analytic Hierarchy Process: Experiences in Chile*).*

Collion, M. -H. and A. Kissi. 1995. Guide to Program Planning and Priority Setting. Research Management Guidelines No. 2. The Hague: International Service for National Agricultural Research.

Guide to an eight-step approach to planning and priority setting within a research program. The approach was pilot tested in Morocco and applied in several countries.

Contant, R. B. and J. A. Bottomley. 1988. Priority Setting in Agricultural Research. Working Paper 10. The Hague: International Service for National Agricultural Research.

Introduction to the need for formal priority setting and overview of a range of tools from which countries and institutions can choose to suit their circumstances. The work was commissioned by the Special Program for African Agricultural Research and supervised by an international team of research leaders.

Contant, R. B. and J. A. Bottomley. 1989. Manual for Methods of Priority Setting in Agricultural Research and their Application. Priority Setting Training Document Version 3.3. The Hague: International Service for National Agricultural Research.

This manual presents examples of priority-setting tools focusing on within-program priority setting based on eight critical factors for use in weighted scoring or benefit/cost analysis, supported by a computer application.

Falconi, C. A. 1998. Methods for setting priorities in agricultural research and biotechnology. In *Managing Biotechnology in a Time of Transition* edited by J. I. Cohen. The Hague: International Service for National Agricultural Research.

Presentation and comparison of different ways to conduct priority setting for biotechnology research: scoring, mathematical programming, simulation, and analytical hierarchy process, with cases.

Franzel, S., J. Jaenicke, and W. Janssen. 1996. Choosing the Right Trees: Setting Priorities for Multipurpose Tree Improvement. Research Report No. 8. The Hague: International Service for National Agricultural Research.

A detailed guideline for a program-level priority-setting exercise aimed at choosing tree species for use in agroforestry.

Janssen, W. 1995 Priority setting as a practical tool for research management. In *Management Issues in National Agricultural Research Systems: Concepts, Instruments, Experiences* edited by M. Bosch and H. -J. A. Preuss. Munster: LT Verlag.

Janssen, W. and A. Kissi. 1997. Planning and Priority Setting for Regional Research: A Practical Approach to Combine Natural Resource Management and Productivity Concerns. Research Management Guidelines No. 4. The Hague: International Service for National Agricultural Research.

These guidelines present a method for within-country regional research programs with a view to combining productivity and natural resource management issues within a regional, decentralized context. The method was pilot tested in Benin and applied in Morocco, Senegal, and elsewhere.

Kamau, M. W., D. W. Kilambya, and B. Mills. 1997. Commodity Program Priority Setting: The Experience of the Kenya Agricultural Research Institute. Briefing Paper No. 34. The Hague: International Service for National Agricultural Research.

Mills, B. 1998. Agricultural Research Priority Setting: Information Investments for Improved Use of Research Resources. The Hague: International Service for National Agricultural Research.

Norton, G. W., P. G. Pardey, and J. M. Alston. 1992. Economic issues in agricultural research priority setting. *American Journal of Agricultural Economics*, 74 (5): 1089–1094.

This article elaborates on economic considerations in priority setting.

Okali C., J. Sumberg, and J. Farrington. 1994. Farmer Participatory Research: Rhetoric and Reality. London: Intermediate Technology Publications.

This handbook gives a realistic view of the possibilities of and approaches to farmer involvement in agricultural research.

Romero, C. and T. Rehman, 1989. Multiple Criteria Analysis for Agricultural Decisions. Amsterdam: Elsevier.

A standard mathematics-based reference for the use of multiple-criteria analytical tools for decision-making in agriculture.

Shumway C. R. and R. J. McCracken. 1975. Use of scoring models in evaluating research programs. *American Journal of Agricultural Economics*, 57 (Nov.): 714–718.

Summary description and critical evaluation of a weighted scoring model for ranking research problems of the North Carolina Agricultural Experiment Station for budget and incremental staff allocation.

Chapter 18
Budgeting

Hilarion Bruneau

A budget compiles the various cost estimates for project funding or a list of identified expected annual revenues and expenses. Budgeting is not a stand-alone process. It follows planning and priority setting and must be supported by sound financial management policies, systems, and practices. A budget is a critical tool for translating plans into research actions. It is thus an integral part of research planning and realization. By following up on planning, budgeting can make researchers' and research managers' jobs easier and help get research done in the most productive way. In many organizations, however, budgeting is used strictly as a control mechanism by top management, funding agencies, or ministerial authorities. To develop a budget, research managers need to be familiar with budgeting systems and mindful of the institutional mission, goals, and objectives. The link with planning is critical. Weak planning and priority setting will lead to weak budgets, limiting research realization. The art is to master the bridging of plans, priority setting, and budgeting, and support research execution with sound financial management practices.

What is budgeting?

Budgeting is "translating the operational short term agricultural research plans into financial terms, so that limited available financial resources can be applied in the most efficient manner to carry out the agricultural research activities described in that plan" (Nickel 1989). As a research management tool, budgeting can serve a number of purposes:

- communicating goals and objectives to staff
- coordinating efforts and activities
- motivating research personnel to meet objectives
- anticipating and avoiding financial problems
- bringing resource allocation in line with priorities

Budgeting in research management

Budgeting is an integral part of the process of research planning and implementation. In developing a budget, research managers focus sharply on the institutional mission and previously established research plans, programs, and priorities. Comparison with and reference to previous planning and priority-setting efforts ensures linkages and coherence of research activities with budgeted and available resources. Multiyear budgeting (budgeting annually over three to five years) is increasingly common (see figure 1). It facilitates the budgeting process from year to year and increases accuracy in budget allocations. It is also useful in forecasting recurrent costs, where it contributes to greater cost awareness on the part of research leaders and partners. When recurrent costs are likely to rise (due to rapid growth in research capacity or spending obligations) it is imperative to forewarn financing agencies of the impending increase in requirements (Tabor, Janssen, and Bruneau 1998).

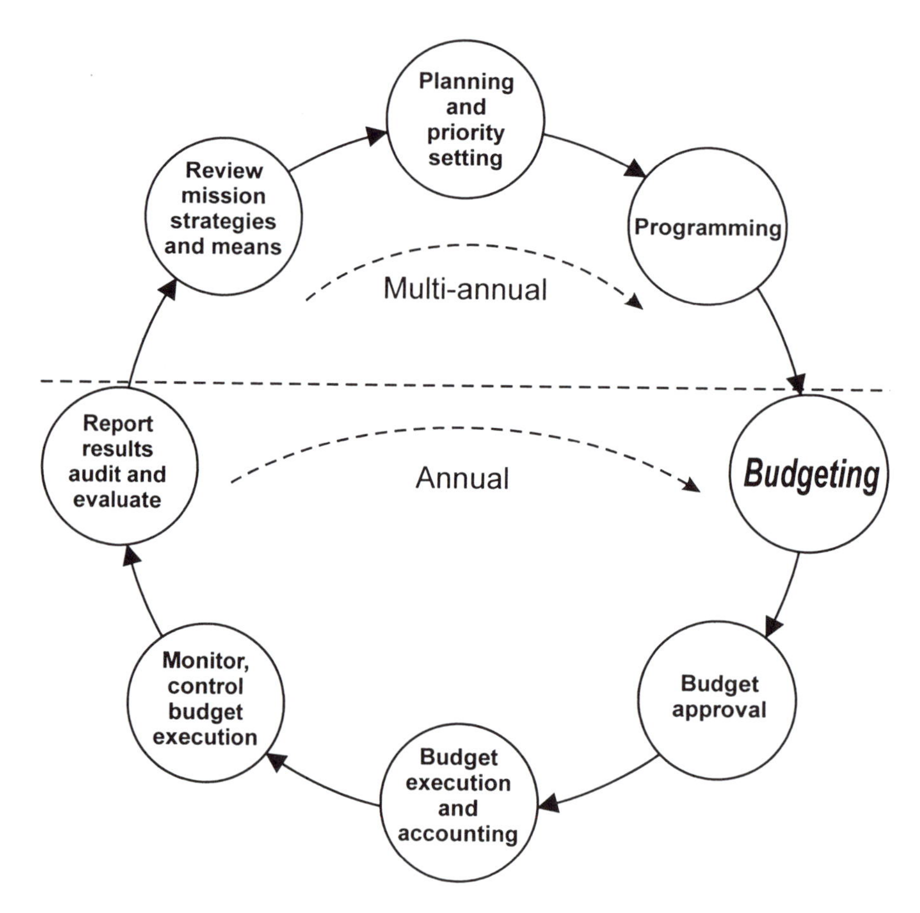

Figure 1. Annual and multiyear budgeting

Budgeting in financial management

Budgeting is an important step in putting money to work to execute agricultural research. It is a management subsystem to help managers and stakeholders plan and manage the deployment of resources effectively and efficiently. It also helps the research system acquire some financial autonomy and facilitates research execution. Because it contributes to increased financial efficiency, good budgeting can take some financial strains off the national budget. Figure 2 depicts the place of budgeting in the financial management process. This process itself contains a number of critical elements:

- appraisal of funding needs and availability (financial planning)
- acquisition and management of funds (financial strategies)
- allocation of funds (*budgeting*)
- use of funds (financial policies, systems, and treasury management)
- control of funds (control, internal auditing, and financial analysis)
- accounting (bookkeeping, gathering financial information, and independent auditing)

What are the roles of budgeting and accounting, and should they be separated? Essentially, accounting is an information and reporting function. It is central to the financial management process. Accounting reports and informs research managers about budget execution. Bookkeeping of budgetary accounts

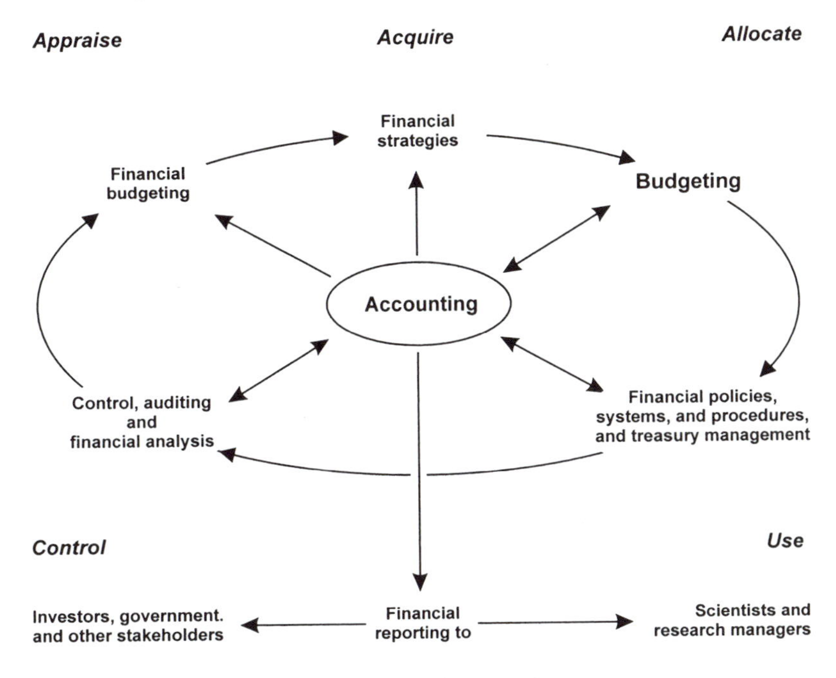

Figure 2 The place of budgeting in the financial management process

belongs in accounting. To avoid biases and conflicts of interest in recording and reporting, accounting and budget preparation (and follow-up) should preferably be separated. Budgeting is closer to planning and programming and critical to getting research done. Many agricultural research organizations, however, lack the resources to separate accounting and budgeting. In such cases, research managers should be aware of possible recording and reporting biases and of a low priority being given to budgeting as a short-term action planning process.

Problems with budgeting

Problems with budgeting in agricultural research organizations have to do with limited knowledge and awareness of the budgeting process as a research management tool, wrong approaches to the development and implementation of budgeting policies and procedures, cash flow problems, public-service traditions, and dependency on external donors. These factors explain why actions to improve budgeting in research organizations have often failed.

Knowledge and awareness

Financial difficulties are often seen as a matter of lack of funds only, rather than as a problem of how limited financial resources are allocated (budgeting). Research organizations may even be unaware that their budgeting process is at fault and that they could alleviate their financial difficulties with better budgeting. Many are reactive rather than proactive in budgeting and rely exclusively on ministerial guidelines. They ask for more resources than they actually require and spend whatever they receive in the hope of getting more next year, rather than adopting a proactive business-like approach (asking what is needed and optimizing the impact of resources received). There is little formal budget preparation consultation and research staff are hardly involved in preparing budgets. The process essentially relies on the fund-raising capabilities of the director and a few key scientists. The following are some practical problems:

- Qualified staff with financial management capabilities are in short supply.
- Budgeting is done largely outside the organization, in government ministries or donor initiated projects.
- No budgeting procedures, standards, or norms are in place; or, where such guidelines exist, there is no written explanation.
- Financial data for budget analysis and preparation are lacking.
- Budgeting and accounting are not integrated. Labels and codes used for budget preparation differ from those used for accounting. This makes budget preparation and comparison of actual and budgeted expenses very difficult.

Often, the budget is a financial wish list, not a daily work tool that matches available funds with the activities that can be realized with the funds available.

Confusion of financial wishes with financial realities leads to misunderstandings between treasury officers, investors, research managers, and scientists.

Approach and methodology

Agricultural research organizations have sometimes sought to solve their financial problems with the simple acquisition of new accounting or budgeting software – a mistake that has proven costly and unproductive. They often underestimate the complexity of their financial problems and the internal resistance to change that finance can engender, as the following common problems illustrate:

- The difficulties and importance of preparing a budget are underestimated. Centralized top-down budgeting generates dependency, resentment, high costs, low morale, misinformation, and actions that cannot be sustained.
- Expenditure is not monitored since this surveillance responsibility is not clearly assigned to persons in charge of budget execution. Financial problems that arise are handled by crisis management.
- Directors keep staff in the dark about their budget. When researchers move to execute previously approved research, they must then request budget information and approval. The approved budget is not communicated to staff. It remains a well guarded secret.
- Either there is no budget calendar, or the one in use is out of sync with the agricultural research cycle. In the latter case, the national budgeting calendar might be used by default, putting the accent on fiscal dependence rather than on research requirements. The budgeting cycle of an agricultural research organization should be synchronized with the agricultural cycle.

Cash flow problems

The reality of national budgets in poor countries is that shortages are frequent in treasury offices, and even for approved budgets cash flow to agricultural research organizations is often interrupted. Yet agricultural research is a seasonal activity, and late or interrupted funding can mean the loss of an experiment and the waste of funds already spent. Good financial management and budgeting can reduce the impact of cash-flow shortages on ongoing research.

Public service traditions

Budgeting practices are adversely affected by traditional public-sector management approaches that emphasize control rather than management. In some research organizations the budget is prepared not based on planned research activities but on the control of inputs by line item. Research organizations in the public sector have a history of reliance on external public structures to handle financial matters. This weakens their own financial responsibility and accountability. For example, the research staff payroll is often handled by a national public-service commission or similar agency, leading to limited financial information for research managers and making the cost of research personnel difficult to monitor.

Donor orientation

In many agricultural research organizations, donor reporting requirements take priority over the development of organization-wide sound financial management practices. The budget is fragmented, consisting of numerous sub-budgets, with no coherence in the time period they cover and presentation of figures. A plethora of research activities and funds are effectively outside the institutional budgetary process and funding channels.

Doing budgeting

The development of a master budget is one of research management's most important tasks. The master budget provides a summary of investments (capital) and recurrent (operational) costs of institutional activities, research, and other activities on the basis of research and administrative units, research programs, projects, and geographical or regional locations. In developing a pragmatic approach to budgeting, managers must first decide whether to follow ministerial instructions or take a more proactive approach and develop a comprehensive and all-inclusive (consolidated) budget of their own. The next step is to select the type of budgeting systems to use.

There are many types of budgeting systems: line item; incremental; formula; the planning, programming, and budget system (PPBS); and the zero-based budgeting system (ZBB), to name but a few. The more complex systems, such as PPBS and ZBB, share some elements with planning and are particularly useful in times of financial scarcity and concern for efficiency and organizational change. Essentially, the tasks of budgeting are four:

- identifying costs and revenue centers (program, project, station, laboratory, technical support, and administrative units)
- appointing a person responsible for the budget of each center or unit (a budget holder)
- allocating revenue and directing costs to the units
- empowering budget holders with the tasks of budgeting, monitoring, and execution

The budgeting approach should be mission-driven, participatory (down and up cycle), decentralized, and result-oriented. Budget meetings should address research in the following order:

- ongoing research (protect previous investments)
- short-term research operations (do what needs to be done now)
- consolidate research results already achieved (capitalize on previous research)
- new research projects and activities (adjust to new demands and priorities)

The inclusion of new projects in short-term budgetary discussions depends on practical considerations of availability of finance for implementation. Consensus on the importance of new projects should have been obtained in earlier research planning exercises. Budget preparation is the time to assess everything being done and incorporate new plans and priorities.

Requirements for good budgeting

Budgeting requires human, physical, and financial resources and competes with other institutional subsystems for those resources. It is closely linked to other research management subsystems. Further, budgeting has its own structure and nomenclature and requires certain skills and know-how. First is "know-how and know who," that is, knowledge of the budgeting process and of the institutional budgeting environment. Second, leadership, commitment, and support are required, leaving no doubt as to top management's determination to master the process and prepare a good budget. A qualified person must be in charge of the budgeting process (a budget coordinator), and back-up is needed from ministerial authorities, treasury officials, donors, and other research partners and stakeholders. Third, linkages to planning and programming are required, mainly consisting of participation of scientists in budgeting. Fourth, funding and facilities are needed to instigate sound budgeting procedures, including money to manage and operate the budgeting process, logistics, office space, equipment, computer services, software, and transport. Finally, tools and information are required. The software used for budget preparation should be simple, widely available, and user friendly (a simple database or spreadsheet). It should have linkages (same account code numbers) with the accounting system and be in accordance with the budget and accounting system of the main source of funding, often a ministry of agriculture or finance.

Relevance for agricultural research

In many agricultural research organizations, despite the time that scientists spend on budget matters, budgeting per se is not given enough attention. Yet the results of poor budgeting practices are visible and form very real obstacles to research execution and impact. With growing competition among public services for scarce financial resources, money to do agricultural research is in short supply. That means its use and impact must be optimized. Research organizations are considered efficient if they make cost-effective use of available resources to produce and transfer appropriate technology to farmers and others involved in agricultural production, processing, and marketing systems. Budgeting is a critical research management tool for optimizing the use and impact of resources. A 1996 World Bank review of achievements and problems in the development of national agricultural research systems recommends that "adequate budget allocations" be a criteria for continued World Bank support (World Bank 1996). Moreover, forecasting and overcoming financial difficulties may depend on good budgeting practices.

Being cost conscious and developing a "budget culture" is particularly important in a volatile financial situation, when cuts are the rule rather than the exception. The first response to insufficient funding should be to manage available funding well. Responsibilities in budget execution should be clearly defined. For an organization to develop an internal budget culture of financial responsibility, budget holders must be identified, empowered, adequately supported by good financial services, and evaluated on their financial results. This means that the costs allocated to budget holders must be under their direct control. This does not mean that budget holders should have easy access to cash or bank accounts. Rather, they are the approving authority for direct research costs in budget execution within the framework of organizational financial management guidelines. Indirect costs and overhead should also be clearly identified, with responsibility delegated to appropriate budget holders.

Examples

This section provides two examples. One relates to the budgeting process and the other to the budget format. Table 1 presents a 12-step procedure for annual budgeting in a public-sector agricultural research organization. The process begins in March, 10 months before the start of the actual budget year (January through December).

The second example (tables 2a, 2b, and 2c) illustrates a budget format at three hierarchical levels: project, program, and organization. At the organizational level both line item and program budget formats are presented. Seeing an example of the actual output of the budgeting process may be of value to researchers, research managers, investors, and stakeholders.

Table 1. Annual Budgeting Process in a Semiautonomous Agricultural Research Organization

Action	Tasks	When	Who
1. Plan	– update budgeting procedure, prepare circulars and forms for next year, distribute these to budget holders – collect estimates of available finances and needs (capital/investments and operations) – organize budget preparation steering committee and plan data collection and meetings for budgeting	March	budget coordinator
2. Coordinate	– budget preparation steering committee meets to decide on and adopt the budget preparation procedure	April	budget coordinator
3. Distribute	– distribute circulars and forms to budget holders – budget holders inform their staff of the ways and means of budget preparation for the coming year	May	budget coordinator and budget holders
4. Prepare	– budget holders collect data within their unit, review and complete budget forms, consolidate data, and forward documents to budget coordinator	June	budget coordinator and budget holders
5. Analyze	– budgeting committee analyzes budget proposals; estimates and justifies proposed expenses and revenues are compared with previous financial and budgetary results; coherence of plans, priorities, and costs are examined – regional meetings are held with budget holders and staff for analysis	July	budget coordinator and budget holders
6. Approve	– revise budget proposal and consolidate information – prepare an operational or activity plan, that is, a proposed budget by activity and timing, by expenses and revenues, and by cost center, research program and/or units and budget holders – prepare detailed analysis of the consolidated budget proposal – approval by budget committee (internal)	August	budget coordinator
7. Present	– present, defend, and negotiate budget proposal before the board of trustees (administrators) and perhaps ministerial authorities (agriculture, finance, or others)	September	top management
8. Adjust	– budget committee meets, discusses, and requests budget holders to adjust proposals and operational plans according to the results of presentations and negotiations.	October	budget coordinator
9. Submit	– Submit for board, ministerial, and parliamentary approval (external)	November	top management
10. Notify	– notify budget holders of the budget approved and request that the information be passed on to staff of their units – input budget data in the accounting system (budgetary accounting)	December	budget coordinator and budget holders
11. Monitor	– execute budget and monitor execution monthly; analyze variances	January through December	budget coordinator and budget holders
12. Amend	– discuss budget execution and variance; arbitrate and modify budget; present and distribute amended budget.	midyear	budget coordinator and budget holders

Table 2a. Consolidated Budget Format for a National Agricultural Research Organization ('000 monetary units)

Traditional – Line Item Budget					**Program Budget**				
	Core	Non-core	Total	%		Core	Non-core	Total	%
Revenue					**Revenue**				
National budget	12.1		12.1	60	National budget	12.1		12.1	60
Donor contracts		5.9	5.9	30	Donor contracts		5.9	5.9	30
Others	2.0		2.0	10	Others	2.0		2.0	10
Total revenue	14.1	5.9	20.0	100	**Total revenue**	14.1	5.9	20.0	100
Expenses					**Expenses**				
Salaries	8.6	0.2	8.8	44	Rice	5.5	0.0	5.5	28
Labor	2.5	1.2	3.7	18	Grain legumes	2.3	1.2	3.5	18
Supplies	0.6	1.5	2.1	10	Oilseeds	1.8	0.0	1.8	9
Travel	0.2	0.9	1.1	6	Fruit	1.0	0.0	1.0	5
Maintenance	0.6	0.0	0.6	3	Vegetables	2.2	0.0	2.2	11
Other	1.6	0.2	1.8	9	Farming systems	1.2	2.8	4.0	20
Subtotal	14.1	4.0	18.1	91	***Subtotal***	14.1	4.0	18.1	91
Capital	0.0	1.9	1.9	9	Capital	0.0	1.9	1.9	9

Source: Adapted from Nestel and Gijsbers 1991, part 1, p. 6.

Table 2b. Grain Legumes Program

Project (experiment)	**Salaries**	**Labor**	**Supplies**	**Travel**	**Maintenance**	**Other**	**Total**
Cowpea germplasm evaluation	660	275	0	180	15	120	1250
Cowpea N-P-K trials	100	40	30	10	0	20	200
Pigeonpea spacing	40	30	20	0	0	10	100
Pigeonpea drought tolerance	250	125	150	30	25	50	630
Soybean breeding	600	330	100	80	60	150	1320
Total	1650	800	300	300	100	350	3500

Table 2c. Cowpea Germplasm Evaluation

Research Actions	**Salaries**	**Labor**	**Supplies**	**Travel**	**Maintenance**	**Other**	**Total**
Soil sampling	125	50	0	0	0	0	175
Staking and tracing	100	40	0	10	0	20	170
Plowing	110	30	0	0	15	10	165
Seeding and fertilizing	150	50	0	30	0	50	280
Thinning	75	50	0	80	0	0	205
Harvesting and sampling	100	55		60	0	0	215
Laboratory costs						20	20
Station costs						20	20
Total	600	275	0	180	15	120	1250

References and recommended reading

Anthony, R. N. and D. W. Young. 1994. Management Control in Non-profit Organizations. Burr Ridge, Illinois: Richard D. Irwin, Inc.

This book is intended for a course on management-control problems in nonprofit organizations in general. While it is not a book on accounting, it does provide insight into financial management tasks from a management control point of view. Chapter 10 is about budgeting.

Byerlee, D. and G. Alex. 1998. Strengthening NARS: Policy Issues and Good Practice. Washington, D.C.: World Bank.

CGIAR Secretariat. 1988 to 1995. Financial guidelines series and related documents. Washington, D.C.: Secretariat of the Consultative Group on International Agricultural Research.

These financial management guidelines were drafted for the international agricultural research centers supported by the CGIAR. Many are applicable to agricultural research organizations in general. Also papers prepared for general distribution in the context of "re-engineering the CGIAR planning, budgeting, and funding systems" describe a funding matrix that can be adapted to developing-country agricultural research systems.

Nestel, B. and G. Gijsbers. 1991. An Overview of INFORM: An Information Management System. INFORM Guidelines 1 and 2. The Hague: International Service for National Agricultural Research.

This volume outlines an information management system developed by ISNAR and aimed specifically for agricultural research managers. It provides insights into the importance and use of financial information in agricultural research organizations for programming and budgeting of research activities.

Nickel, J. L. 1989. Research Management for Development: Open Letter to a New Agricultural Research Director. San José, Costa Rica: Instituto Interamericano de Cooperación para la Agricultura.

This slim book is a frank and lucid account of numerous aspects of agricultural research management. Chapter V is about budgeting and managing funds.

Osborne, D. and T. Gaebler. 1993. Reinventing Government: How the Entrepreneurial Spirit is Transforming the Public Sector. Reading: Addison-Wesley P. C., Inc.

This book is about introducing entrepreneurship into public service institutions and increasing their efficiency. It emphasizes the need to change and a new approach to financial management in public institutions.

Menard, L. 1994. Dictionnaire de la comptabilité et de la gestion financière: anglais-français avec index français-anglais. Ordre des experts comptables (France); Institut des réviseurs d'entreprises (Belgique). Montreal: Institut Canadien des Comptables Agréés.

This dictionary defines more than 8000 terms used in financial management based mainly on North American and European practices. It is nonetheless applicable internationally. The dictionary is English-French with a French-English index.

Tabor S., W. Janssen, and H. Bruneau. 1998. Financing Agricultural Research: A Sourcebook. The Hague: International Service for National Agricultural Research.

World Bank. 1996. Achievements and Problems in Development of NARS. Report No. 15828. Washington, D.C.: World Bank.

Chapter 19
Implementation

Jaime Tola, Govert Gijsbers, and Helen Hambly Odame

There is a growing awareness of the importance of developing an implementation-oriented plan. Traditional planning approaches assumed that implementation was merely an administrative activity undertaken by operational staff under instructions from decision makers higher up in the organization. Now implementation is seen as a difficult task, but one that can be facilitated by actions taken early in the planning process. Without this forethought even the best designed agricultural research plans will be worth little more than the paper they are written on. Implementation is a complex, political process that requires leadership and management, as well as feedback through careful monitoring and evaluation.

What is implementation?

Strictly speaking, implementation is beyond planning. But implementation is so essential to give practical shape to the plans developed – to ensure that plans are actually fulfilled by concrete measures – that a sourcebook on planning cannot overlook implementation concerns. Moreover, implementation issues often condition the scope and feasibility of research plans. If implementation is not considered in the planning process, the effectiveness of the plan will most certainly be jeopardized.

Plan implementation has received much less attention than planning itself. This is remarkable, because planning, in a way, can be seen as a preparatory phase to ensure that activities, projects, programs, and policies are carried out in the most effective manner. But theory and advice are largely missing when it comes to implementation. Neglect of implementation can be explained by the fact that for a long time it was seen as a straightforward administrative task.

As with planning, the different dimensions of implementation may be discussed under the headings of "context," "process," and "content" (Grindle 1980, Najam 1995). Analyzing and dealing with the *context*, or environment, in which implementation takes place is now recognized as a significant management issue. Several external factors help explain why implementation is difficult. First, planning often implicitly assumes a stable context in which, once designed and approved, plans will automatically be implemented. The real world, however, does not stand still while an organization readies itself to implement plans. So rather than being a straightforward, linear process that is periodically monitored using predetermined "milestones" and indicators, changes in the external environment require that plans be adjusted to "catch up" with events.

Second, implementation is not a set of neutral, administrative activities. It is highly political. This is particularly so in agricultural research and development, in which a large number of stakeholders, often with conflicting objectives and agendas, are involved and held accountable for plan implementation.

Third, implementation is not always the first objective of the planning unit or the committee charged to develop the plan. Plans are sometimes designed for other, mainly external reasons including administrative requirements, donor funding, and public relations.

From the *process* perspective, planning and implementation have usually been, respectively, responsibilities of planning units and operational units. Yet weak or absent links between planners and implementers make plans difficult to realize. The discrepancies between what is planned and what can be implemented may be so large that the plan loses the credibility and ownership it attempted to build in the planning process. The position, roles, and responsibilities of planners and their interactions with program staff responsible for implementation are thus critical issues.

Mintzberg (1994) discusses the links between planning and implementation, concluding "the blame has to be laid, neither on plan formulation nor on implementation, but on *the very separation of the two*" (italics original).

It used to be assumed that implementation would happen automatically after a plan had been approved. This idea of "machine-like" implementation effectively conceptualized implementation as a set of administrative tasks and, as a result, minimized the importance of implementation. In the 1970s, authors such as Wildavsky used an empirical approach to analyze implementation. Based on extensive case studies, they concluded that implementation is highly complex and that no set rules apply. These studies were criticized for being "atheoretical," case-specific, and overly pessimistic as to the potential for successful implementation. However, they did demonstrate that even good policy or plans can and do go wrong in implementation.

Others have attempted to develop a theory of implementation or, at least, a set of guiding principles (Morah 1990). Here, concern is less with specific implementation failures and more with trying to understand how implementation processes work in general. These attempts, however, tend to debate implementation in terms of top-down and bottom-up models. More recently, interactive models of implementation have emphasized a synthesis between top-down and bottom-up approaches.

The *content* or substance of implementation is a function of both the type and the scope of the plan that is being implemented. With regard to the type of plan, Crosby (1996) distinguishes a continuum of three main types: policy implementation, program implementation, and project implementation (table 1). At the research policy level, implementation is strategic in nature, while at the program and project levels, it is more operational.

Implementation issues also reflect the scope and complexity of research programs and projects. Challenged to increase the relevance of their work, agricultural research organizations take on increasingly complex research subjects, such as natural resource management and agroindustrial concerns – areas in which implementation capacity has yet to be developed.

How plans affect implementation can be described along three dimensions. First, the plan and its implementation may focus narrowly on research activities or they may include additional stages in the technology transfer and innovation process. Each stage involves different actors, partnerships, and resources. Second, programs and projects may be implemented at a single location or in a number of different agroecological zones at different times and in various research organizations. Coordination between organizations in different regions is costly and management-intensive. Third, some plans focus on research objectives exclusively, while others explicitly incorporate organizational or national development objectives. These additional aims add implementation challenges.

Doing implementation

There can be no fixed rules or solutions for problems in implementing agricultural research plans. At best the study and practice of implementation offers only a set of "guiding principles":

1. *Avoid and reduce administrative control problems.* The implementation of plans should take into consideration the characteristics of the implementing

Table 1. A Continuum of Implementation Tasks

Policy Implementation (emphasis on strategic tasks)	**Program Implementation**	**Project Implementation (emphasis on operational tasks)**
– legitimation – constituency building – resource accumulation – organizational design/structure – resource mobilization – monitoring impact	– program design – capacity building for implementers – collaboration with multiple organizations and groups – expanding resources and support – proactive leadership	– clear objectives – defined roles and responsibilities – plans/schedules – rewards/sanctions – feedback/adaptation mechanisms

Source: Crosby 1996.

organization and power relations that will influence decisions and actions in plan delivery. Most importantly, plan implementation may imply changes in administrative routines, and staff might resist such changes. Agreements made during the planning process must be reinvigorated at the implementation stage to ensure that a supportive environment flourishes within the organization.

2. *Ensure adequate resources.* Early consideration of the type and amount of human, financial, and technical resources required for implementation is an essential part of the planning stage.
3. *Resolve disagreements over goals.* The implementation of agricultural research plans often involves different stakeholders and implies the reconciliation of diverse cultural, economic, and political interests. Agreement on the goals of the plan is fundamental to fostering commitment to the implementation process.
4. *Encourage dialogue and pressure for change from the target group.* Plans will not be adopted or will eventually break down without sustained interest from the target population.
5. *Facilitate complex joint action.* The larger the number of partners involved, the greater the flexibility required in the implementation process.
6. *Leadership is essential.* Implementation benefits from strong leadership that sustains and unites those who are committed to the successful delivery of a particular plan. Bourgeois and Brodwin (1998) emphasize the chief executive's role in linking planning and implementation as a major condition for success.
7. *Reduce uncertainties by monitoring and evaluating during implementation.* Implementation of plans may be somewhat unpredictable, but progress towards goals should be carefully monitored.

Relevance for agricultural research

The importance of implementation seems so obvious that it hardly seems to merit attention. Yet while most research organizations formulate new plans, they give scant attention to delivery on those plans. Implementation is not a set of routine administrative tasks, but a complex, often politically charged process, particularly in agricultural research programs characterized by multiple stakeholders pursuing a variety of interests. Much goes wrong between formulating the plan and achieving its objectives. Organizations that are aware of the different implementation problems can take corrective action. This section discusses what can go wrong by means six implementation "pathways" as presented in figure 1.

In each subfigure (*a* through *f*) of figure 1, the circle on the left shows the plan at the start. The right-hand circle indicates the end of the implementation

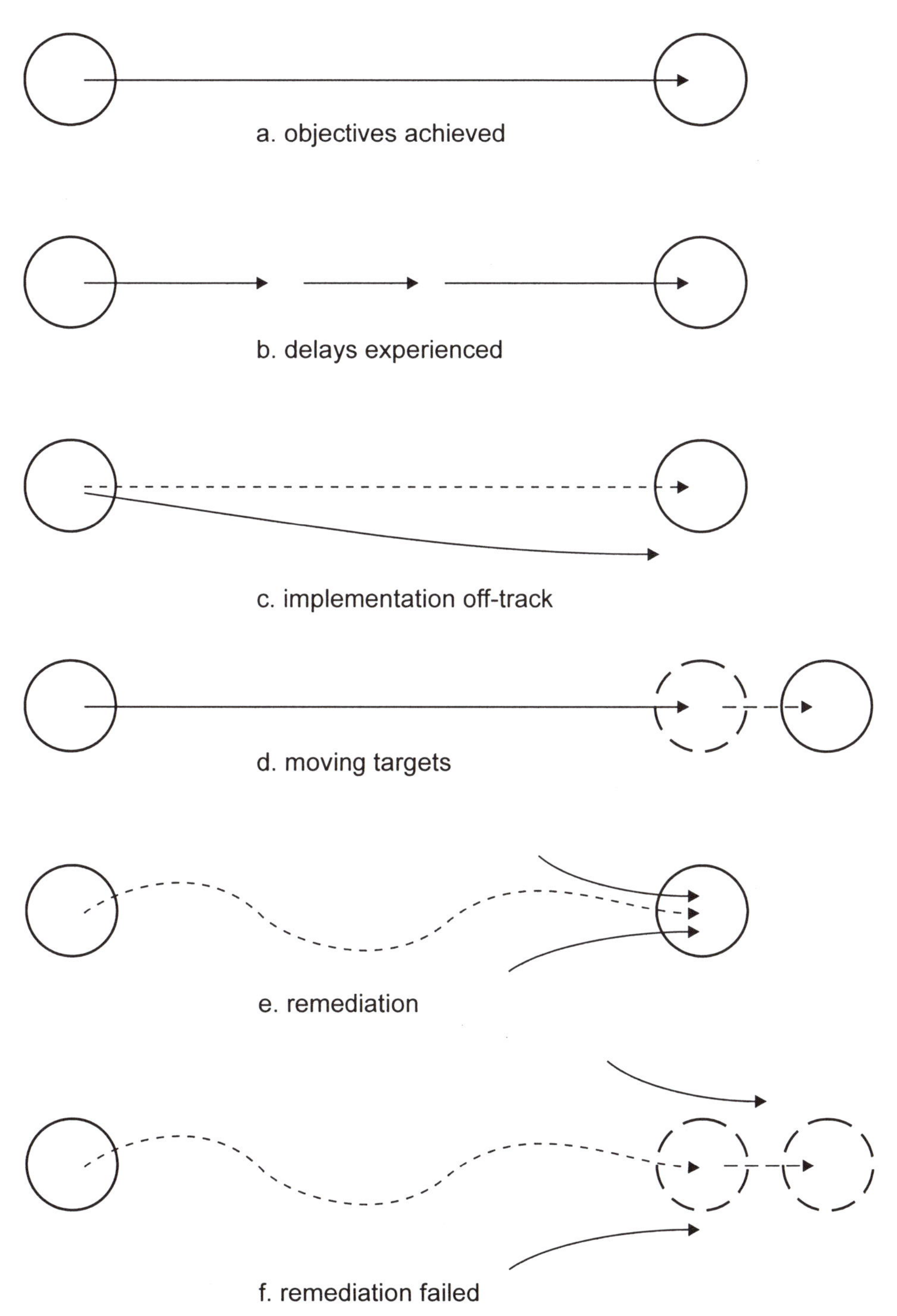

Figure 1. Implementation pathways

process. Figure 1*a* shows the rather unusual situation in which implementation proceeds entirely as foreseen. The circle on the left is a solid plan that is implemented without problem and the plan's objectives are achieved on time. A simple variation is shown in figure 1*b*, in which the plan's objectives are achieved, but not without delay. Delays may have severe implications for the achievement of the plan's aims, for example, by jeopardizing funding.

Figure 1*c* depicts a situation where both the plan and its objectives are solid and valid, but the implementation process goes off track; the target is missed and the objectives are not, or only partially, achieved. This is a common situation caused by problems in the implementation process itself. Funding may not materialize as promised, trained staff may leave the organization, and internal conflicts may derail the process.

Figure 1*d* shows a more complex situation in which the original plan was relevant to the problem situation. But during the implementation process things changed either inside or outside the organization to such an extent that the original targets are no longer relevant. The target has moved. This situation may occur in the face of external shocks: policy changes may require different research objectives or a devaluation may make a range of technologies unprofitable for farmers.

These situations are similar in that all were based on "good" plans at the starting point. This is, however, not always the case. Plans may be weak to start with for a number of reasons: data on and knowledge of the current situation may be limited; the plan may represent an uneasy compromise between the interests of different stakeholders; the plan document may have been drawn up mainly to attract external funding, while internal commitment to it was weak; or the plan's objectives may be too ambitious. Such situations will not necessarily lead to disaster. Weak plans can sometimes be salvaged. A remedy is possible if there is sufficient political will and leadership, if external conditions develop in a favorable direction, or when additional resources become available. This is shown in figure 1*e*. Finally, figure 1*f* presents a situation where failure is complete, with no possibility for remediation. The plan as originally conceived has major problems, the target has moved, and resources and commitment are weak. Essentially this requires planners to start again.

The six implementation pathways help us to understand that implementation can go wrong in many different ways and for many different reasons. It also helps us to understand that smooth implementation is the exception rather than the rule. Most plans do get in trouble or need revisions during implementation. But with strong leadership and early attention, these problems can often be remedied.

Examples

Many agricultural research organizations are currently undergoing a transition. Such organizational reform processes often involve deep changes. In general, all change processes – and the planning documents that embody them – are motivated and enforced by reorientation on socioeconomic conditions and the need to respond to complex and new technological demands. Sometimes these processes are partially or fully implemented; others exist only in written proposals and plans. Nevertheless, most agricultural research organizations have the strong desire to fill the gap between what they are and what they need to be in order to contribute effectively and efficiently to agricultural development. The following two cases from research organizations in Latin America briefly reflect some tasks and challenges encountered in plan implementation.

Ecuador

The agricultural research sector of Ecuador underwent a process of diagnosis and planning that led to the creation of the legally autonomous Instituto Nacional Autónomo de Investigaciones Agropecuarias (INIAP). This process spanned nine years (1987–95), with planning documents available in 1992. Implementation of the plan began in the subsequent four years. Elements included strategic planning, upgrading salaries and salary-scale policy, and significant reduction and decentralization of the research program structure.

These reforms, together with a government endowment for INIAP, provided 65% of the institute's annual budget. Despite these achievements, an important setback in implementation was the lengthy process of negotiating loans with international donors for the Ecuadorian agricultural sector. These delays directly affected scientific resource development and the realization of INIAP's strategic plan.

Uruguay

Uruguay's Instituto Nacional de Investigación Agropecuaria (INIA) is a case of building an entirely new, decentralized research body complete with a public-private matching budget and equivalent governance structure. The process started in 1989, and the new, autonomous INIA emerged in 1995. As in Ecuador, the process of institutional change was assisted by ISNAR, using an approach known as "diagnosis-planning-implementation." Implementation was left in the hands of the national organizations. Six main lessons were learned:

1. Any process of redesigning an agricultural research institution is lengthy and complex.
2. To realize organizational changes, a national policy for agriculture research must create a legal framework supportive of the change process.

3. To institutionalize basic organizational change, training is required to curb resistance at management and operational levels; changes require participation of all staff.
4. Institution building based on isolated or unsystematic management actions has little impact and can adversely affect staff morale and performance.
5. Ownership and institutionalization of changes play key roles in successful implementation.
6. A skillful combination of strategies for dealing with staffing and human resource development improves the effectiveness of implementation.

Finally, the Uruguay case showed that it was less important to specify implementation tasks than to consider early-on how the process of implementation should proceed. The institute maintained an interactive approach that emphasized ownership by the national elements and a synthesis between top-down leadership and bottom-up participation. This has helped INIA maintain relevant, effective, and efficient agricultural research.

References

Bourgeois, L. J. and D. Brodwin. 1998. Linking planning and implementation. In *Strategy: Process, Content, Context* edited by B. de Wit and R. Meijer. London: International Thompson Business Press.

Brinkerhoff, D. W. 1996a. Process perspectives on policy change: highlighting implementation. *World Development,* 24 (9): 1395–1401.

Brinkerhoff, D. W. 1996b. Coordination issues in policy implementation networks: An illustration from Madagascar's environmental action plan. *World Development,* 24 (9): 1497–1510.

Crosby, B. L. 1996. Policy implementation: The organizational challenge. *World Development*, 24 (9): 1403–1416.

Grindle, M. S. 1980. Politics and Policy Implementation in the Third World. Princeton: Princeton University Press.

Mintzberg, H. 1994. The Rise and Fall of Strategic Planning. Englewood Cliffs, N.J.: Prentice Hall, Inc.

Morah, E. U. 1990. Why Policies Have Problems Achieving Optimal Results: A Review o f the Literature on Policy Implementation. UBC Planning Papers, Discussion Paper number 20. Vancouver: University of British Columbia.

Najam, A. 1995. Learning from the Literature on Policy Implementation: A Synthesis Perspective. Laxenburg, Austria: International Institute for Applied Systems Analysis.

Chapter 20
Towards an Integrated Planning, Monitoring, and Evaluation System

Douglas Horton and Luis Dupleich

Monitoring and evaluation go hand in hand with planning. Planning is essential for effective monitoring and evaluation, and lessons learned from monitoring and evaluation enable better planning in the future. Integrated planning, monitoring, and evaluation (PM&E) improves governance, decision making, learning, and overall performance of agricultural research organizations. It helps organizations reduce duplication of efforts and paperwork. A process for strengthening and integrating PM&E includes the following steps: (1) assessing current PM&E procedures, (2) envisioning an "ideal" PM&E system that the organization should work toward over time, (3) developing an action plan for strengthening and integrating PM&E, (4) implementing the plan and monitoring progress, and (5) periodically reviewing results and revising plans and implementation strategies.

What is an integrated PM&E system?

An integrated PM&E system is to an organization what the central nervous system is to a living organism. It helps each part keep in touch with the other parts and with the external environment. It orients behavior and coordinates actions in the pursuit of common goals. It provides information for learning lessons and improving performance. There is no "blueprint" for PM&E. The challenge is to apply general principles, as outlined below, to craft an integrated PM&E system using a variety of the available building blocks.

Linked components

At the level of the organization, strategic planning and external reviews are carried out about once every five years. Internal reviews are organized annually. An annual report is published yearly as well, and impacts of programs and selected activities are assessed intermittently. At department or program level, research is planned and reviewed each three to five years. Operating plans, budgets, and program reports are prepared on a yearly basis. At the level of the project, research activities have their own specific objectives, budgets, and time frames. Procedures are in place for project preparation, review, approval, supervision, budgeting, reporting, and evaluation. These procedures are synchronized with the organization's annual budgeting and reporting cycles. Facts and figures on research inputs, activities, and outputs may be compiled in a project-based information system, which may aggregate project-level information to produce program and department-level reports, as well as reports on specific research activities and results for the organization as a whole.

PM&E activities, such as those outlined above, are recorded in internal documents that records results, decisions, actions required, and actions taken within the organization. In addition to formal PM&E activities, informal ones, such as technical seminars and field visits facilitate communication between researchers and managers, to align their efforts in pursuit of common goals and to promote organizational learning.

Throughout the management cycle

PM&E and impact assessment can be viewed as stages in a management cycle (figure 1). The management-cycle concept, widely applied in project management (Horton et al. 1993), can also be used at other decision-making levels, including the program, the research center, and the organization as a whole. At the start of the cycle, planning incorporates assessments of client needs and research opportunities, definition of goals, and setting priorities. Plans include targets and milestones that serve as reference points for subsequent monitoring and evaluation. Later in the cycle, activities are implemented and progress monitored in relation to plans. Activities and outputs are recorded for use in reporting and evaluation. Insights gained through monitoring are used to improve plan implementation. At the end of the cycle – or earlier if warranted – an evaluation is conducted to assess the quality, efficiency, and effectiveness of the project. After research results have been made available to potential users, an impact study may be carried out to assess the benefits – and the negative effects – of new knowledge or technology.

Across decision-making levels

Researchers and managers working at different levels within the organization require different types of information because they make different types of decisions. Whereas top management may set broad organizational goals, researchers may have considerable autonomy in planning and implementing specific research activities. In an integrated PM&E system, PM&E activities carried out at each decision-making level are consistent with, and informed by, those carried out at other levels. For example, a farmer consultation carried out in one region provides an input to the organization's strategic planning. Later in the

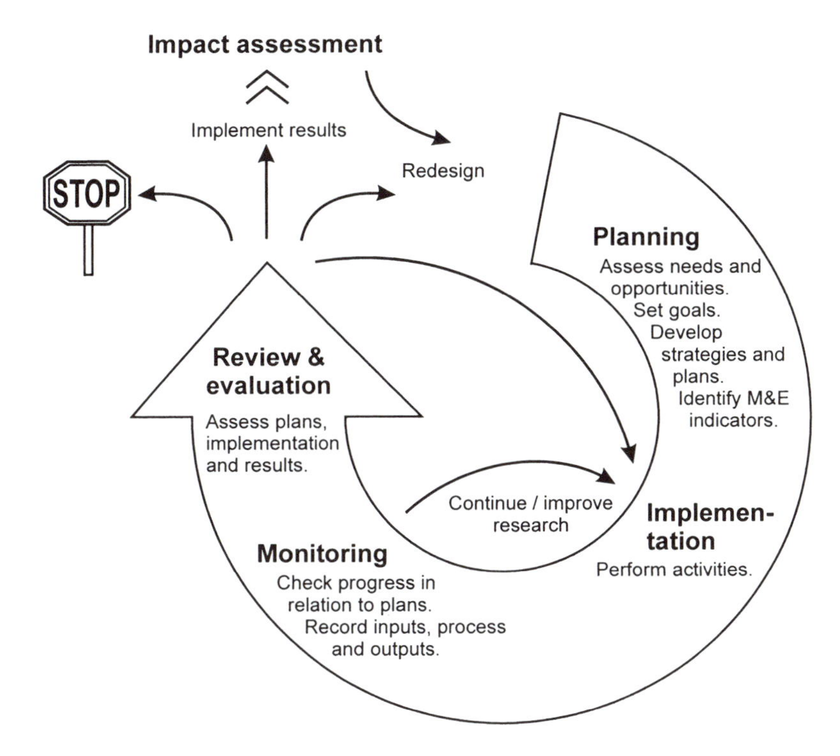

Figure 1. The planning, monitoring, and evaluation cycle (based on Horton, 1998, figure 3)

management cycle, project-level progress reports provide basic information for preparing program and organization-wide reports.

Administrative procedures

Research organizations often have two separate management systems: one for managing research activities and the other for managing research staff, facilities, and other resources. The first system is often thought of as research management and the second as administration. While these two spheres are clearly related, the procedures and individuals responsible for them are often separate. Consequently, researchers may be burdened with two different sets of procedures. A fully integrated PM&E system seamlessly connects procedures for research program management with those of administration.

Compatible procedures

Within an agricultural research organization or system, units working in different geographical areas or on different technical subjects may require different management cycles and procedures. For example, vegetable research trials may be planned and evaluated every few months, but trials involving cattle or perennial crops may last several years. Nevertheless, integration of PM&E requires a minimum set of standardized procedures to be used, so that information from different units can be assembled, aggregated, and applied in organization-wide analysis, decision making, and reporting.

Clear responsibilities and scheduling

Managers and researchers at each decision-making level and in each operational unit need to know what PM&E tasks they are responsible for, when they are to be performed, and what information is to be produced by whom and in what form. Overall coordination may be provided by directors of research or programs or by one or more specialized management units (for example, a PM&E unit or separate planning and monitoring and evaluation units). In an integrated system, PM&E activities are coordinated and synchronized across organizational units, to ensure that the information required for decision making is available in an appropriate form and at the time it is needed.

Internal and external stakeholders

Integration within the PM&E system is insufficient, however. To serve as a strategic management tool, an institutional PM&E system must involve both internal and external stakeholders, so their interests and concerns are reflected in the resulting plans and reports.

Developing an integrated PM&E system

An integrated PM&E system must be tailored to meet the specific needs of the organization. To operate on an ongoing basis, it must be embedded in the organization's structure and culture. It is inadvisable to try designing a fully integrated PM&E system and then "install" it in an agricultural research organization. Nor should one attempt to "transfer" a successful PM&E system from one organization to another, without carefully assessing the degree to which it fits the needs and circumstances of the recipient organization. The best course is for each agricultural research organization to develop and integrate its own PM&E system gradually, over time. Based on experiences in strengthening PM&E and managing change in a number of organizations, a five-step approach has been suggested for developing an integrated PM&E system (Horton et al. 2000).

Step 1. Assess current PM&E procedures

The first step in developing an integrated PM&E system is to assess current PM&E procedures in relation to the needs and expectations of the organization's members and external stakeholders. Three tasks are required for such an assessment:

1. *Identify the needs, concerns, and expectations of key stakeholder groups.* There are four key stakeholder groups within the organization: top managers, PM&E technical staff, middle managers, and researchers. External stakeholders include the organization's clients, partners, and funding agencies. For each group, three questions can be asked: "What types of information are desired, when and in what form?" "How would stakeholders like to be involved in PM&E?" And, "What are their personal assessments of current PM&E procedures?"
2. *Document current PM&E procedures.* The main PM&E activities currently undertaken need to be described: their purpose, who is responsible, what resources are used, what methods are employed, what outputs are produced, and how outputs are used. What problems have been encountered? And, to what extent have they been overcome? To what extent is PM&E integrated? Do plans include milestones and indicators? Are evaluation results used in future planning?
3. *Identify strengths and weaknesses of current procedures in relation to the identified needs and expectations.* This task requires stakeholders' involvement, perhaps at a workshop. Based on results of the first and second tasks, the main strengths and weaknesses of current PM&E can be identified. Priorities for improving procedures can then be set.

Step 2. Envision an ideal PM&E system

The second step is to envision a PM&E system that efficiently meets the needs of staff and external stakeholders. A well structured vision is composed of the *values* on which the PM&E system will be developed, a *goal* for PM&E, and a vivid *description* of what it will be like to accomplish these goals. Collins and Porras (1996) argue that for an organization to achieve enduring success it needs a clearly defined purpose and set of guiding values, which together constitute its vision. This idea also applies to the development of a PM&E system. The types of PM&E activities to be carried out also should be described, along with responsibilities for PM&E, resources to be employed, outputs to be produced and benefits expected from the PM&E system. Operational details and procedures are not required at this point. As said, an organization-wide PM&E system cannot simply be designed and installed – it needs to be developed over time. A fully integrated system that meets all the requirements identified above is not a feasible short-term goal. Rather, it represents an ideal to aspire to and work towards.

Step 3. Prepare an action plan

The next step is to develop an action plan for moving from the current situation to the preferred scenario. Once the organization has defined its preferred PM&E system, an action plan is needed to guide the organization from the present to the preferred state. The action plan identifies a sequence of activities to be implemented and for each activity, *when* the activity will be carried out, *who* will be responsible, *what* precisely will be done, what *resources* will be employed, and what *results* will be produced

Step 4. Implement the plan, monitor progress, and make adjustments

Steps 1 to 3 prepared the way for activities that will improve PM&E. Thorough preparation eases the implementation process and facilitates learning and continuous improvement of the PM&E system. However, without vigorous implementation, time and energy spent in planning will have been wasted. Step 4 incorporates implementing the plan, monitoring progress, and making necessary course corrections.

Step 5. Review results and revise goals and strategies

The PM&E system should be reviewed periodically to ascertain the relevance of its objectives and to check the design of the system, the efficiency and effectiveness of implementation, results to date, and expected future benefits. Over time, changes in the stakeholder environment, in technology available for PM&E (including information technology), in the organization's strategy or structure, or in other factors may call for a redesign of the whole PM&E system or of individual procedures.

Managing the change process

Implementation of the steps outlined above constitute a process of organizational analysis and change that requires careful management. Experience learns that the management can best be provided by a "change team" that reports to top-level management. The change team is primarily responsible for analysis, development of proposals, and implementation of agreements. Top management is responsible for deciding on key issues throughout the change process.

Relevance for agricultural research

Benefits of integrated PM&E systems in agricultural research are manifold. First and foremost, when PM&E is integrated in a management cycle, each phase in the cycle is improved. Efforts made during planning to develop a hierarchy of objectives and assumptions, to identify expected outputs, and define milestones and indicators for later monitoring and evaluation improve planning because they tend to make plans more realistic and down-to-earth. Similarly, monitoring is improved when it is done both to identify divergences from plans and to gather information useful for evaluation. And when projects are assessed in relation to prior plans, based on systematic data gathered during implementation, and with the purpose of improving subsequent cycles of activities, the evaluation is likely to be focused, based on information (rather than evaluators' prejudices), constructive, and of direct use to stakeholders.

Second, systematic monitoring and evaluation of activities in relation to plans encourages the preparation of more realistic and coordinated plans and the implementation of planned work. A problem often noted in agricultural research organizations is that individual researchers work in isolation rather than as members of a team. A common result is the proliferation of small, dispersed research projects or activities that may contribute little to the achievement of organizational goals. Similarly, information on research activities and outputs is often dispersed in individuals' files or in decentralized programs. Such scattering limits the organization's ability to focus resources on priority problems and to build an organizational memory that enables staff to marshal information when needed.

Third, an integrated PM&E system helps organizations respond more effectively to external demands. Stakeholder participation and systematic procedures for PM&E provide valuable tools for focusing organizations' activities on topics of importance to external groups and for responding to changing needs and concerns as they arise.

Fourth, integrating PM&E in a management cycle fosters individual and organizational learning and may strengthen staff motivation. While discrete PM&E activities are valuable in and of themselves, benefits for individual and organizational learning are far greater when PM&E is carried out as interrelated phases within a management cycle. Embedding PM&E in an organization's operations and culture greatly facilitates organizational learning. Recent studies also indicate that researchers support and participate actively in PM&E when they helped design the procedures and feel that such activities contribute to research programs and results (Horton and Mbabu 1998). By the same token, staff tend to resist PM&E if it imposes bureaucratic demands or evaluation with little scientific basis and is not used in decision making (Rajeswari 1999).

Finally, information produced by a PM&E system is useful in preparing institutional presentations. Organizations are increasingly challenged to present their goals, activities, and results coherently and convincingly. Such presentations are essential in fund-raising, project proposals, and reports to donors and stakeholders. As demands for such presentations proliferate, institutions need to draw more on institutional PM&E systems that retain information on projects and programs in readily accessible form.

Potential costs and problems

PM&E entails a few potential costs and problems too. The major cost of a PM&E system is the time scientists and managers need to design and implement the system. Where the organization has several sources of funding, managers will need to design a system that meets both external and internal management requirements, while placing minimal demand on scientists' time and energy. A poorly managed PM&E system can lead to "mindless paperwork." Care must also be taken to determine just what are decision makers' information needs. Otherwise, much information may be collected that is never used. To avoid this situation, which is especially common in monitoring systems, managers should weigh the value of specific types of information against the difficulty of collecting and processing it. Only information that has a clearly demonstrable value should be collected.

A second potential problem is summed up in the proverb, "The perfect is often the worst enemy of the good." The ideal PM&E system is a goal to work toward. But such a goal can never be achieved if the first steps – however imperfect – are never taken. One promising strategy is to begin with a relatively simple system with robust, "loosely coupled" components. Such a system is less vulnerable to shocks and stresses (such as lack of resources at critical moments) than a highly sophisticated system that may collapse if any one of its components fails.

Examples

Since its establishment in the 1970s, the CGIAR has undergone a continuous process of integrating planning with monitoring and evaluation activities (for more detail on PM&E in the CGIAR, see Ozgediz 1999). As the CGIAR grew and evolved from three to 16 research centers, it developed an increasingly complex PM&E system aimed to promote internal coherence, improve management, and ensure accountability of the research centers to the group's sponsors and clients (national and regional research and development entities).

Initially each center developed its own plans and budgets, using its own approaches and formats. Similarly, each donor evaluated the activities it supported at each center using its own evaluation procedures. In the 1970s, the CGIAR developed procedures and guidelines for budgeting, to facilitate the preparation of a consolidated budget for the system of research centers. Procedures were established for external quinquennial reviews of each center's programs. Centers also began conducting annual internal program reviews.

Since the mid-1980s, the CGIAR's Technical Advisory Committee (TAC) has prepared systemwide priorities and strategies to aid donors in allocating funds among center programs. The priority and strategy documents are updated about every five years. In the mid-1980s, centers prepared their first strategic plans. Since then, they too have been updated about every five years. Also, during the 1980s, the scope of center reviews broadened to cover both program and management issues.

In the 1990s, the CGIAR introduced a matrix management system, in which contributors annually approve a set of program activities that is to be implemented by the research centers. All the centers are now required to group their activities in projects that fall within one of these approved CGIAR "programs." In 1995, an independent impact assessment and evaluation function was established for the CGIAR. Currently, a logical framework approach is being introduced to facilitate project-level PM&E.

Up to now, different groups have been responsible for preparing systemwide priorities and strategies, coordinating external reviews of centers, and undertaking system-level impact assessment. However, in response to a 1998 review of the CGIAR system, responsibilities for these three functions are being merged within TAC.

Cuba's national agricultural science and technology system

In the early 1990s, Cuba's Ministry of Agriculture perceived a need for radical change in the science and technology institutes serving the agricultural sector. It decided to create a national agricultural science and technology system (SINCITA) embracing the 17 agricultural research institutes currently operating in the country.

The first stage in the process involved formulating guiding principles for the system's design and implementation and defining an overall strategy for the process as a whole (a "macro-strategy"). In the second stage, strategic plans were elaborated for the system and for each of the 17 institutes. The third stage included the redesign of the research and development model focusing on external demands and a new PM&E system. The new model defined planning, implementation, validation, and diffusion as key stages in the research and development process. Stakeholder participation was viewed as crucial for identifying research priorities, validating products and services, and diffusing innovation. Demands from Cuban agribusiness were carefully studied as well.

The integrated PM&E system that was established defined three management levels: strategic (system level), tactical (program and subprogram level), and operational (project and subproject level). The "project" was conceived as the basic unit of research management. PM&E was viewed as a means of continuously improving internal processes, as well as research products. Alongside validation of the new PM&E system, a comprehensive manual of norms and procedures for its implementation was elaborated. These guidelines articulate how research and development activities are organized under the Ministry of Agriculture. They describe institutional responsibilities for research and articulates and regulate means for keeping stride with the changing external environment (through advisory boards and expert consultations).

References

Collins, J. C., and J. I. Porras. 1996. Building your company's vision. *Harvard Business Review,* 74 (5): 65–77.

Horton, D. 1998. Disciplinary roots and branches of evaluation: Some lessons from agricultural research. *Knowledge and Policy*, 10 (4): 31–66.

Horton, D. 1999. Building capacity in planning, monitoring and evaluation: Lessons from the field. *Knowledge, Technology, & Policy,* 11 (4): 152–188.

Horton, D., R. Mackay, A. Andersen, and L. Dupleich. 2000. Evaluation of Capacity Development in Planning, Monitoring, and Evaluation: A Case from Agricultural Research. Research Report No. 17. The Hague: International Service for National Agricultural Research.

Horton, D., et al. (eds). 1993. *Monitoring and Evaluating Agricultural Research: A Sourcebook.* Wallingford, UK: CAB International.

Horton, D. and A. N. Mbabu. 1998. Experiences with Research Planning, Monitoring and Evaluation in Kenya: Lessons for Africa. Discussion paper No. 98-10. The Hague: International Service for National Agricultural Research.

Mato, M. A., et al. 1999. Auto-análisis de la consolidación del Sistema de Ciencia e Innovación Tecnológica Agraria (SINCITA) del MINAG de Cuba: Experiencias, lecciones e impactos de un proceso de cambio institucional. Havana: Ministerio de la Agricultura.

Morgan, G. 1998. Images of Organization: The Executive Edition. Thousand Oaks, Calif.: Sage.

Ozgediz, S. Evaluating research institutions: Lessons from the CGIAR. *Knowledge, Technology & Policy,* 11 (4): 97–113.

Rajeswari, S. 1999. Professionalisation and evaluation: The case of Indian agricultural research. *Knowledge, Technology, & Policy,* 11 (4): 69–96.

Part IV
Tools and Instruments for Agricultural Research Planning

Willem Janssen

Introduction

Numerous planning tools are available to help planning teams understand the future environment of an organization and how it might respond to challenges and opportunities that may arise. Planning tools contribute to the planning process in two dimensions. First, they support the planning process by laying out clear procedures and expected outputs and by defining rules for communication. The future is too multifaceted to be fully understood by a single person, and within planning teams different perspectives on the future may exist. Through the use of tools, different personal qualities in developing responses may be combined in a structured manner to arrive at balanced decisions with broader support than would be possible if single individuals were to undertake the planning effort. Second, they enable planners to integrate external information in the planning process. There are many sources of relevant information and many analytical perspectives on what the future may bring. Tools with which to process this information into a form that is readily understood by many people improve the quality of the resulting plan.

This overview discusses how and why planning processes may be supported and their information content strengthened. Afterwards it offers a glimpse of nine tools that are of particular relevance to agricultural research planning. These tools are the subjects of the following chapters.

Supporting the planning process

Bringing people and perspectives together in the planning process is one key responsibility of the planner or planning unit. Most of the people involved will not and should not be professional planners. Their central interest in the planning process is how it affects their work, their role, or their living conditions. In order to use participants' time as effectively as possible, tools are useful for structuring and organizing planning. These tools provide guidance in the planning task, help to define the objective of the planning process (or a step in it), and help people organize their thinking. Tools to support planning processes have often arisen from trial and error in the actual development of plans. Most, therefore reflect a certain judgment of what is or is not a good plan.

Sometimes a series of tools constitutes a complete planning procedure. For example, the German technical assistance organization, GTZ, developed a series of planning-by-objective procedures that they use in planning most of their projects (GTZ 1988). The long-term program planning procedure reported in this book by Collion and the technology foresighting process described by Rutten also apply a series of planning tools towards an outcome. Such procedures may be especially useful when the nature of the planning exercise is well known and agreed upon. If this is not the case, it may be difficult and undesirable to follow a standard procedure. The challenge then becomes choosing and combining tools to support the planning process in the best manner possible. It may even be impossible to fully anticipate which tools will be required.

Tools to support planning tend to concentrate on facilitating communication or on outlining the planning process. To facilitate communication, some simple tools may greatly contribute to the effectiveness of a planning group. Personal introductions before starting to work together are useful, especially if people talk about why they are part of the planning team and what they hope to contribute. Brainstorming is a well known approach to gather numerous ideas, which are organized afterwards. The "metaplan method" is among the best known tools for improving communication (DSE 1985). It facilitates understanding and recollection of the discussion by the use of cards on which central thoughts are expressed. The metaplan method is used in the constraint-tree approach.

Tools such as the analytic hierarchy process have been developed to help decision makers evaluate alternative options. Scoring methods are similar but less formal. Decision support tools are most effective if they are based on a simple evaluation mechanism and lead to easy-to-understand results. Whereas these methods are effective for guiding the decision process, the risk is that they stress subjective (personal) over objective (evidence-based) knowledge.

With the advent of information technology, computer-based tools for supporting planning have gained prominence. An advanced tool is the "decision room." Here, a number of people use a decision support method such as analytical hierarchy process or scoring and evaluate alternatives, immediately discussing agreements and disagreements. Project planning and management software, such as Microsoft Project, may be used to assign responsibilities, fix schedules, and define financial requirements over time. "GANNT charts," which are used to map activities and time throughout execution of a plan, are commonly included. "CPM" (critical path management) or "PERT" (program evaluation and review technique) may also be used. These tools are used to optimize planning activities in highly complex projects so as to be completed in a minimum of time. Where such software appears too complex, responsibility charts may be useful. A simple responsibility chart might indicate the task, persons involved, and the time when a planned activity should be undertaken. Computer software planning tools tend to be more appropriate for planning activities with a short time horizon and a limited scope (e.g., project planning), rather than for planning with a long time horizon and a wide scope (e.g., strategic planning).

Increasing the information content of the planning process

Another set of planning tools serves to integrate more and higher quality information in the planning process. These tools can improve planning in at least three ways:

1. They may help to predict the future environment of the organization and therefore allow for plans that better anticipate changes that may occur.
2. They may improve predictions of expected impacts of possible plans, thereby improving the evaluation of alternative plans.
3. They may improve estimates of resource requirements and resource availability, thereby increasing plan feasibility.

Tools that improve the information base of the planning process have often originated in a specific scientific discipline. For example, cost-benefit analysis, grounded in economics, contributes to understanding expected impacts of alternate plans in the light of resource requirements. Geographic information systems (GIS), from geography, help planners understand the spatial impacts of new technologies.

The most useful tools for improving the information base are those that are easily understood by all people involved in the planning exercise. The maps that can be produced using GIS are easy for most people to understand and interpret. In this respect GIS is an excellent example of a planning tool that produces highly accessible outcomes. There is an almost infinite number of tools that can be used to improve the information base in planning. Planning, however, may suffer from too much as well as too little data. If information to improve planning comes from many different sources, reflects many kinds of parameters, or is based on a wide range of methodologies, the planning group may be overwhelmed and lost. One discipline (perhaps the one with the most representatives in the planning group) may then start to dominate the planning process. Tools to improve the information base should focus on the most essential dimensions of the decision problem, or they should allow participants to truly combine more information towards a relevant plan.

The chapters

The chapters in this part of the *Planning Sourcebook* present some valuable tools for improving research planning. Although many tools contribute to both the dimensions referred to earlier, normally they contribute more to one than to the other. Figure 1 positions the nine tools described in the following chapters on these two dimensions. It suggests that more process-oriented tools (e.g., the logical framework analysis) might be best combined with tools that are more information based (e.g., geographic information systems).

Table 1 overviews some features of these tools, indicating the organizational levels at which they may be used and describing the skills required for successful application. As explained, these are certainly not the only tools that can be usefully applied in a planning exercise. Other tools may be introduced depending on the nature of the planning process.

Using the *analytic hierarchy process* (AHP), complex decisions that are difficult to oversee are broken into smaller parts that are more easily handled. A general goal for the decision-making exercise is defined, and the goal is decomposed into a number of criteria. An assessment is made as to how alternative research programs or projects contribute to the different criteria. This is done by comparing pairs of alternatives with regard to each criterion. The assessments on each criterion are then combined into an overall assessment.

Constraint trees are used to develop a shared understanding of the problems and opportunities that applied research must address in a specific domain, such as a commodity or region. It is an excellent tool for guiding the diagnostic phase

Figure 1. Contributions of planning tools to the information content and the planning process

Key:

GIS	geographic information systems	SC	scenario analysis
SIM	simulation models	CT	constraint trees
MIS	management information systems	AHP	analytic hierarchy process
GA	gender analysis	LFA	logical framework analysis
PRA	participatory rural appraisal		

Table 1. Usefulness and Requirements of Different Planning Tools

Planning Tool	Organizational Level Where Applied	Required Practitioners
Analytic hierarchy process	institute, program	facilitator, management specialist
Constrain- tree analysis	program, project	facilitator, subject matter specialist
Gender analysis	program, project	social scientist
Geographic information system	institute, program	geographer, database manager
Logical framework	project, program	management specialist
Management information system	institute, program, project	management specialist, database manager
Participatory rural appraisal	program, project	social scientist
Scenario analysis	system, institute	economist, social scientist
Simulation models	project	modeling expert

of the planning process, because it allows many different people to participate and many ideas to be reconciled.

By distinguishing the different problems and opportunities faced by men and women, gender analysis allows research programs to target their interventions with more precision to users. *Gender analysis* is typically used to strengthen the diagnostic phase of planning. It is most useful at the research program and project levels.

Geographic information systems (GIS) help planning participants understand the spatial pattern of problems and opportunities in agriculture and to predict the spatial impact of research. This is useful for evaluating the expected success and impact of research strategies. GIS enables results to be presented in attractive maps that are easy for most people to interpret.

Logical framework analysis has become the standard for summarizing a project or program plan. The logical framework describes the causality relating activities to outputs, objectives, and the final goal of a project. It also identifies mechanisms for verifying progress towards outputs and objectives, as well as the critical assumptions that must fulfilled for the project to be successful.

A *management information system* (MIS) includes many elements outside of planning. But it also supports planning processes by providing information on resource use and availability and by feeding results of planning exercises (at the program or project level) back into the information system for monitoring.

Participatory rural appraisal (PRA) may strengthen the information base for planning as well as affect planning procedures. Within the realm of PRA there are many tools for collecting information from farmers and other rural inhabitants. In PRA both information extraction and deciding how to use the information are participatory. PRA thus brings research planning closer to the user.

Scenario analysis helps planning participants view the future without prejudices that may arise from past experiences. If there is considerable uncertainty about the future and the implications of alternative decisions are very significant, scenario analysis may help improve decision making.

Finally, *simulation models* are used to evaluate the impact of alternative plans on the research subject. By using all the information available on a subject and complementing this with additional experiments, models can be constructed that allow prediction of behavior under a series of assumed conditions. Especially for complex subjects (for example, livestock) simulation models may clarify the attractiveness of alternative research strategies and may help planners choose the optimum one.

References

DSE. 1985. Participatory Approaches for Cooperative Group Events: Introduction and Examples of Application. Peissenberg: Deutsche Stiftung für Internationale Entwicklung.

GTZ. 1988. ZOPP: An Introduction to the Method. Eschborn, Germany: Deutsche Gesellschaft für Technische Zusammenarbeit.

Chapter 21
Analytic Hierarchy Process

Thomas Braunschweig

The analytic hierarchy process (AHP) is a decision-support tool for complex multicriteria decision problems. By decomposing a problem into a hierarchical structure, AHP helps decision makers cope with complexity. The weights of the decision criteria and the priorities of the research alternatives are determined by comparing two elements at a time, verbally expressing the intensity of preference for one element over the other. By this pairwise comparison process, AHP enables incorporation of both qualitative and quantitative aspects of the decision problem. Research managers can use AHP to prioritize projects or programs as a basis for resource allocation. The analytic nature of AHP provides a clear rationale for the choices being made. Its conceptual simplicity and intuitive logic facilitate the participation of various stakeholder groups. The rigorous structure of AHP models improves collective thinking, reasoning, and the efficiency of group decision making.

What is the analytic hierarchy process?

AHP is a decision-support tool to tackle complex multicriteria problems, such as selecting priority research projects or allocating research resources. The method helps practitioners structure and analyze a decision problem by breaking it down into a hierarchic order and employing pairwise comparisons of elements to determine preferences among the alternatives. AHP was developed in the late 1970s by T. L. Saaty (1980). It has since been applied to a wide range of decision problems in the public and private sectors (Zahedi 1986, Vargas 1990). The essential components of AHP are the creative process of constructing and analyzing a hierarchy and the analytical process of making judgments. The former provides detailed insight and helps participants achieve a common understanding of the important factors in a decision problem. The latter offers a sound technique for eliciting and quantifying decision makers' preferences. Therefore, AHP is a powerful and flexible approach to decision making that provides the necessary logical/scientific foundations without ignoring the fact that solving complex decision problems is a process that involves creative thinking, learning, and revising of the outcome.

Figure 1 presents a basic hierarchy, which is made up of three levels. The top level is the general goal of the exercise as agreed upon by the participants, for example, "selecting research projects that contribute most to sustainable agricultural development of the country." The second level consists of decision criteria that are conceived relevant for the achievement of the goal. Ultimately, the criteria correspond to national development objectives. But usually they have to be detailed further in order to give them operational meaning. The bottom level encompasses the alternatives: for example, research projects that resulted from a constraint-tree analysis (see Kissi, this volume). The research projects are compared as to how they satisfy each criterion; and the criteria are compared as to how they contribute to the general goal.

By structuring the decision problem according to its component parts and arranging them in different levels, AHP enables decision makers to focus on smaller sets of the problem. It deals with both qualitative and quantitative aspects of a problem, since participants make value judgments in the pairwise comparison procedure based on experience, intuition, and expertise, as well as on hard facts. The approach, therefore, explicitly allows subjective judgments, recognizing their legitimate role in ex ante analysis. This feature is essential, given that research managers often work with incomplete information in planning.

The rigorous structure of AHP models improves collective thinking, reasoning, and efficiency of group decision making. The simplicity and intuitive logic of the method further facilitates participation of diverse stakeholders in the process. A unique feature of AHP is the possibility of measuring how inconsistent decision makers were in making judgments in the pairwise comparisons. This inconsistency measure brings errors to light and may point out the need to revise judgments, thus improving the quality of the decision outcome.

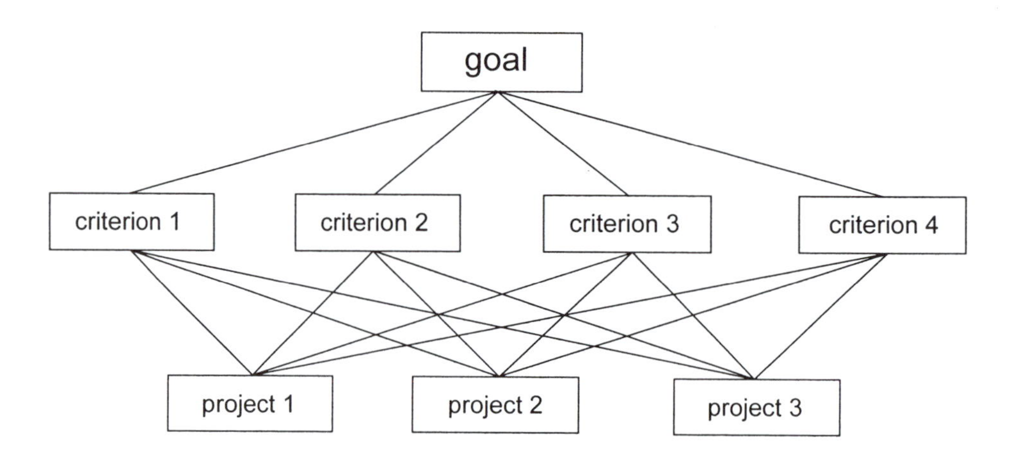

Figure 1. A basic hierarchy

AHP can also be combined with other management tools such as cost-benefit analysis, to determine the economic impact of research activities, or linear programming, to optimize resource allocation. Finally, a software package named Expert Choice© considerably facilitates the application of AHP and is available at reasonable cost (a trial version of the software can be downloaded free of charge from http://www.expertchoice.com/download).

Using AHP

AHP is based on three principles: decomposition of a complex unstructured problem, comparative judgments about the problem's components, and synthesis of priorities derived from the judgments.

Decomposition of a complex unstructured problem

The first step is to break down the decision problem into a hierarchical structure (as in figure 1). Dissecting the problem into its essential components and structuring the components hierarchically helps research managers focus on smaller parts of the decision problem which are easier to handle. The structuring process also improves understanding of the problem since each component (goal, criteria, and alternatives) must be carefully defined. To introduce more precision in the evaluation process, criteria can be split into subcriteria, adding an additional level to the hierarchy. A criterion, for example, "environmental impact," is further specified by breaking it into subcriteria such as "impact on groundwater quality," "impact on soil erosion," and "impact on biodiversity."

Comparative judgments about the problem components

In this second step, research alternatives are evaluated and the criteria weighted. Projects are compared in pairs to assess their relative strengths with respect to each criterion at the next higher level. Similarly, the criteria are compared in pairs to determine their importance with respect to the goal. Table 1 presents verbal terms in the fundamental scale used to assess the intensity of preference between the two elements. The ratio scale and the use of verbal comparisons facilitate the weighting of criteria, as well as evaluation of projects according to nonquantifiable criteria. Once the verbal judgments are made, they are translated into numbers by means of the fundamental scale.

As an illustration, consider three projects A, B, and C, which are ranked with respect to the environmental criterion. First, projects A and B are compared using the questions, "Which of the two projects, A or B, is preferable as judged by the environmental criterion?" "How preferable is it?" The judgment (e.g., project A is very strongly preferred over project B) is then translated into the corresponding numerical value (7) using the fundamental scale and entered into the

Table 1. The Fundamental Scale for Comparative Judgments

Numerical Values	Verbal Terms
1	equally important, likely, or preferred
3	moderately more important, likely, or preferred
5	strongly more important, likely, or preferred
7	very strongly important, likely, or preferred
9	extremely more important, likely, or preferred
2, 4, 6, 8	intermediate values to reflect compromise

matrix shown in figure 2. The next question is on projects A and C; and finally, B and C are compared. For obvious reasons, the cells in the diagonal always have the value 1. Only the judgments on one side of the diagonal have to be elicited since the comparison of project A with project B is the reciprocal value of the comparison of B with A (note that the reciprocal of 7 is $^1/_7$). For each criterion, the projects under evaluation are compared and the judgments entered in a separate matrix.

Synthesis of priorities derived from the judgments

To calculate the priority ranking of the projects with respect to the environmental criterion (the so-called "local priorities") the values in the cells of any column are normalized. That is, they are divided by the sum of the corresponding column. The matrix in figure 3 shows the column sums and the local priorities calculated on the basis of the first column's sum (1/1.48 = 0.68; 0.14/1.48 = 0.10; 0.33/1.48 = 0.22). Expressed in percentages the ranking of the three projects with respect to the environmental criterion is A (68%), C (22%), and B (10%). However, since the matrix of judgments is inconsistent (i.e., if project A is preferred 7 times more than project B and 3 times more than project C, the comparison of project C with project B should yield 7/3 = 2.33 and not 2), the local priorities will differ depending on which column is chosen. To deal with such inconsistencies, the software program mentioned above applies a mathematical process ("eigenvector" method) that uses all of the information contained in the matrix, not just in one particular column. It

Environmental Criterion	Project A	Project B	Project C
Project A	1	7	3
Project B	1/7	1	1/2
Project C	1/3	2	1

Figure 2. Matrix to elicit pairwise comparisons

also enables calculation of a measure of decision maker's inconsistency in making judgments.

Since several criteria are usually applied in evaluating research projects, the local priorities must be combined taking the criteria weights into account. This synthesis is done by multiplying the relative priorities of each project by the corresponding criteria weight and adding them to yield the final composite priorities with respect to the goal stated at the top level of the hierarchy. Often it is desirable to test the stability of the ranking to changes in the criteria weights. Sensitivity analysis can be performed for this purpose, using different weights based on scenarios depicting alternative future developments or diverging views on the relative importance of the criteria. The final ranking of the research projects and the outcome of sensitivity analysis can then be used for allocation decisions.

Relevance for agricultural research

AHP can be used by agricultural research leaders to select research projects or programs. Such programming and funding decisions are fundamental to agricultural research planning. Furthermore, a shift is underway in the mode of research funding, from institutional budget assignment towards project-based funding. This is due partly to calls for heightened research efficiency and client orientation. In particular, competitive grants have become popular in many countries' agricultural research systems, and AHP provides a useful tool for supporting allocation decisions in such schemes. Choices among projects typically involve determination of trade-offs among competing objectives. Such multicriteria decision problems become increasingly complex as the agricultural research agenda broadens, incorporating objectives of non-traditional stakeholder groups. Moreover, demands for accountability and cutbacks in public research budgets suggest the need for more systematic priority setting. AHP is a formal approach that helps elicit, categorize, order, compare, and summarize information and data systematically. Thus, the approach provides clear rationale for a particular choice. This transparency improves communication and facilitates acceptance of the results.

Environmental Criterion	Project A	Project B	Project C	Local Priorities
Project A	1	7	3	0.68
Project B	1/7	1	1/2	0.10
Project C	1/3	2	1	0.22
Column sum	1.48	10	4.5	

Figure 3. Matrix with local project priorities

Information is key in priority setting. However, analysts are often faced with a weak information base, because relevant secondary data is unavailable and collection of primary data too expensive. The problem of data availability leads analysts to rely on subjective judgments to generate information on likely costs, benefits, and other variables connected with alternative research activities. By explicitly recognizing and incorporating subjective judgments, the AHP-based framework encourages participants to pool their knowledge and expertise.

Transparency is critical when subjective judgments are elicited. Participants provide more accurate information when they clearly understand what is expected of them and the procedure to be followed in the exercise. Moreover, communicating the reasoning behind decisions and the procedure by which decisions were made requires opening the "black box" of priority setting. Agricultural research projects sometimes deal with sensitive issues of public interest; they always involve public resources. Successful implementation of the research projects chosen therefore depends on broad acceptance of the decisions made. For this, a transparent priority-setting process that is easily communicated is a precondition. AHP improves understanding of complex decisions and communication of the outcome by dissecting the problem into a hierarchical structure and assigning intensities of preferences among pairs of elements. Its systematic and comprehensible procedure facilitates, in particular, communication with stakeholders not directly involved in the process.

Since the AHP approach focuses on the *process* of priority setting, it helps the participants learn to discern what makes a project valuable. Participation of various stakeholder groups also promotes consensus building and strengthens ownership of the decision outcome, which builds momentum for implementation.

The flexibility of AHP in structuring decision problems and its procedure of relative comparisons facilitate the incorporation and evaluation of research-specific and often intangible variables such as the potential for success, probability of technology adoption, or contribution to building capacity. Similarly, its flexibility permits research planners and managers to use it to tackle other relevant decision problems, for example, the selection of a candidate for a job, the choice of the most appropriate laboratory equipment, or deciding between alternative sites for a new experiment station.

AHP is not, of course, without shortcomings. The main one is the heavy workload involved in pairwise comparisons in cases with a large number of alternatives. Whereas a set of four research projects requires six judgments per criterion, for eight projects 28 judgments have to be made for each criterion. One way out of this problem is to use a "rating mode," that is, absolute (as opposed to relative) measurement of the alternatives. A second potential shortcoming concerns the risk of oversimplifying the decision problem to save time. Since there are no clear limits for where to stop the disaggregation process, the proper application of AHP requires some experience. Furthermore, there is the risk of over reliance on subjective judgments when other information might be available that could inject more objectivity into the decision process.

Because AHP is so flexible, it can be adapted to almost any planning budget. The costs of applying AHP can be managed in three ways: (1) the number of criteria and subcriteria can be varied, (2) participation can vary from a single program analyst to a large group of stakeholders, (3) criteria assessments can be based on subjective information or on collected information. One reason why AHP has become popular is that managers can apply it as a quick individual brain exercise or as a tool to improve collective reasoning and decision making. Playing with these factors has, of course, implications in terms of the quality of reasoning, ownership, and empirical grounding. However, AHP is not more costly than other priority-setting methodologies with comparable detail of analysis.

Example

Assisted by ISNAR and the Swiss Federal Institute of Technology, Chile's national agricultural research institute (INIA) applied AHP to prioritize research projects to be funded under its national biotechnology program (Braunschweig and Janssen 1998, Braunschweig et al. 1999). Two groups participated in the exercise. The strategic group, consisting of research leaders and policymakers, defined and weighted the decision criteria. The technical group, with the project leaders and representatives from INIA's planning unit, assisted in the structuring process and evaluated the research proposals. Because the strategic group faced time constraints, their judgments were elicited in individual interviews, whereas the technical group gathered in two workshops, both facilitated by a moderator. The result of the exercise was a structured list of weighted decision criteria, a priority ranking of the evaluated projects, and a set of scenarios reflecting the different criteria weights.

Decomposing the problem

The first task was to hierarchically structure the decision problem. For this purpose, the general goal of the exercise had to be explicitly stated, a list of relevant criteria established and agreed upon, and the structure of the hierarchy developed. In addition, each criterion in the hierarchy was carefully defined and indicators identified that captured their meaning. The elements of the hierarchy and its basic structure were discussed with the strategic group. In a workshop, the technical group then elaborated the necessary details. Since the participants also evaluated the potential for research and adoption success of each project by means of AHP, two specific hierarchies capturing the potential impacts of the projects were constructed in addition to the main hierarchy. The combined outcome of the hierarchies yielded the final project priorities. Only the latter is presented here (figure 4) as illustration. Note that the hierarchy consists of a clearly defined goal, a level with the four main criteria derived from national development objectives, another level with subcriteria further specifying the criteria above, and a lowest level comprising the research projects to be evaluated (named after the crop to which they relate).

Comparative judgments about the components

In the second step, members of the strategic group weighted each criterion and subcriterion individually. The technical group evaluated the biotechnology projects with respect to each subcriterion. Both the weighting and the evaluation were performed using the pairwise comparison procedure. The matrix to elicit the weights of the main criteria is presented in figure 5. It shows the judgments provided by one member of the strategic group. The criteria weights employed for the calculation of the final project priorities were averages taken from the weights provided by the individual group members.

Synthesis of priorities derived from judgments

After the participants performed the pairwise comparisons for the criteria, subcriteria, and research projects, the weights and local priorities were calculated using the mathematical process integrated in the Expert Choice software. The average criteria weights are shown in figure 6. Since the weights of the main criteria varied markedly among members of the strategic group, sensitivity anal-

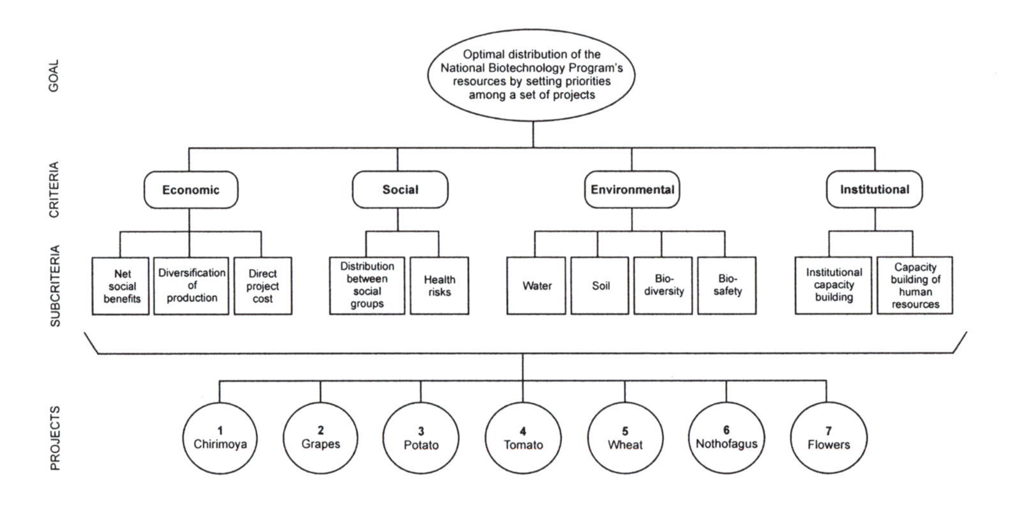

Figure 4. Hierarchy to evaluate the potential impact of the research projects

With Respect to the Goal	Economic	Social	Environmental	Institutional
Economic	1	3	3	4
Social	1/3	1	1	1/3
Environmental	1/3	1	1	1
Institutional	1/4	3	3	1

Figure 5. Judgments for the weighting of the main criteria

yses using different weightings were performed to test the stability of the final project ranking.

Finally, the weights and priorities from the impact hierarchy were synthesized and combined with those from the assessment of the probability of success in order to arrive at the final ranking of the biotechnology projects (figure 7).

The AHP-based priority-setting exercise was judged favorably in its application in Chile. The personal and institutional commitment from the Chilean side contributed much to its success. Various information sources, including extensive subjective judgments, were tapped to obtain the information needed. The decision to work with a strategic and a technical group was correct in this case, given the different kinds of judgments that were expected from the experts. Sensible project priorities together with other results of the exercise provide a strong basis for resource allocation. For Chile, the most important finding was the wide variation in the experts' weighting of the criteria. Those responsible for the national biotechnology program plan to follow up on this. Estimations of the time and cost of the AHP exercise and resource requirements for future applications show the cost to be well in line with the recommendations of international organizations for priority setting in agricultural research.

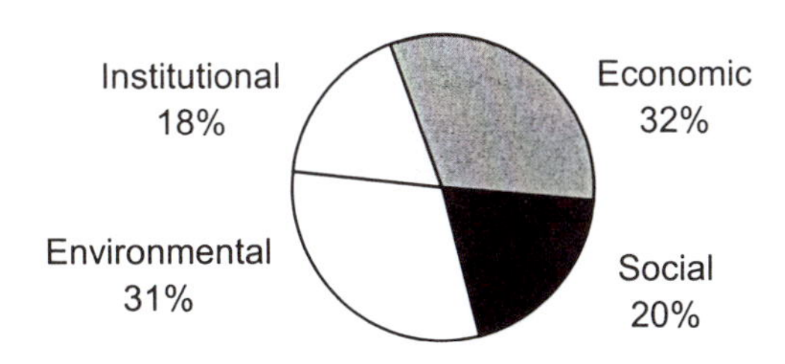

Figure 6. Average weights for the main decision criteria

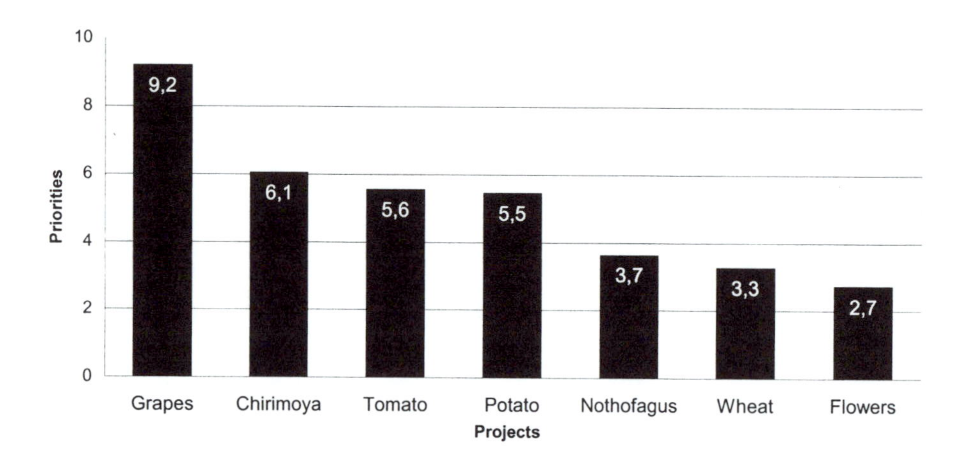

Figure 7. Final project rankings

References

Braunschweig, T. and W. Janssen 1998. Establecimiento de prioridades en la investigación biotecnológica mediante el proceso jerárquico analítico. Research Report No. 14. The Hague: International Service for National Agricultural Research.

Braunschweig, T., W. Janssen, C. Muñoz, and P. Rieder. 1999. Setting research priorities for the Chilean biotechnology program. In *Managing Agricultural Biotechnology: Addressing Research Program Needs and Policy Implications for Developing Countries* edited by J. I. Cohen. Wallingford, UK: CAB International.

Dyer, R. F. and E. H. Forman. 1992. Group decision support with the analytic hierarchy process. *Decision Support Systems,* 8: 199–124.

Golden, B. L., E. A. Wasil, and P. T. Harker (eds.) 1989. The Analytic Hierarchy Process: Applications and Studies. New York: Springer Verlag.

Lockett, G., B. Hetherington, P. Yallup, M. Stratford, and B. Cox. 1986. Modelling a research portfolio using AHP: A group decision process. *R&D Management,* 16 (2): 151–160.

Saaty, T. L. 1980. The Analytic Hierarchy Process: Planning, Priority Setting, Resource Allocation. New York: McGraw-Hill.

Saaty, T. L. 1994. Fundamentals of Decision Making and Priority Theory With the Analytic Hierarchy Process. Pittsburgh, Penn.: RWS Publications.

Saaty, T. L. 1995. Decision Making for Leaders: The Analytic Hierarchy Process for Decisions in a Complex World (third edition). Pittsburgh, Penn.: RWS Publications.

Saaty, T. L. and L. G. Vargas. 1994. Decision Making in Economic, Political, Social and Technological Environments With the Analytic Hierarchy Process. Pittsburgh, Penn.: RWS Publications.

Vargas, L. G. 1990. An overview of the analytic hierarchy process and its applications. *European Journal of Operational Research,* 48 (1): 2–8.

Zahedi, F. 1986. The analytic hierarchy process – A survey of the method and its applications. *Interfaces,* 16 (4): 96–108.

Website

http://www.expertchoice.com/
For information and a free trial version of the Expert Choice software for supporting AHP.

Chapter 22
Use of Constraint Trees in Research Planning

Ali Kissi

Agricultural research planning must be based on a collective understanding of the factors that curtail the performance of the sector, or parts of it. Factors limiting performance are often called "constraints." The "constraint-tree approach" is a means of identifying the most important constraints facing the sector. The constraint tree is usually constructed in a workshop. The first step in building the tree is determining the single, major constraint. This constraint synthesizes all the factors affecting the performance of the (sub)sector. The causes of this major constraint are identified in the next step, after which the causes of each of the identified causes are determined, and so on. Once all the constraints are identified, specific research projects can then be developed to overcome them. The constraint tree is presented as a flow chart of boxes placed at different levels. The constraint in each box results from constraints in the boxes below.

What is a constraint tree?

Planning a research program in any given agricultural sector or domain (e.g., a commodity system, group of commodities, ecosystem, or agroecological region) should be based on a solid and shared understanding of the technological requirements for improved performance. These requirements usually correspond to environmental, technical, and socioeconomic factors that limit productivity and market potential or cause degradation of the resources employed.

A constraint tree is a tool for the systematic identification and analysis of the factors that affect the development of a given agricultural production sector. It is built around a single, major constraint that encompasses and synthesizes all factors limiting the development of that sector's potentialities as well as all factors that cause or aggravate the sector's degradation.

The constraint tree is presented in the form of a flow chart with several boxes placed at different levels. Each box corresponds to a constraint. The constraint in each box results from constraints in the boxes below, and so on. Thus, level 1, or the "treetop," represents the major constraint. Level 2 presents the immediate causes of the major constraint and comprises the main branches of the constraint tree. Boxes in level 3 show the causes of the constraints identified at level 2 and make up the secondary branches, and so on. In such a chart, cause-and-effect relationships are indicated by a system of "arrows" pointing from cause to effect (see figure 1).

The number of primary branches varies according to the sector or domain, but is usually between three a (pointed tree) and seven (a spread and ramified tree). The number of branches and levels tends to be large when the tree is constructed for a national commodity or regional program or for a research program that includes several commodities. For a national program, some of the branches may describe agricultural production constraints in different regions or production systems. For example, in Morocco's olive research program, three of the six branches depict the causes of decreased productivity in the three main production systems: irrigated, intensive, and extensive.

The primary branches may also reflect technical (production, harvest, conservation) or socioeconomic constraints or constraints related to natural resource management. Table 1 presents examples of the different categories of branches in the constraint trees of selected research programs in Morocco and Benin. Branches may encompass technical, natural resource, and socioeconomic aspects. Thus, in the branch "declining yields in the Saiss region" (see figure 1) nontechnical constraints are included such as "lack of incentives."

Building constraint trees

Constraint trees are best developed in participatory planning workshops by a group of people representing different interests or positions. To improve the information basis for the building the tree, a (sub)sector analysis may be undertaken in preparation for the workshop. The scope and depth of the sector study depends on the complexity of the sector under consideration and the time and human, physical, and financial resources available. Sector studies are often undertaken by multidisciplinary teams and are based on bibliographic reviews and primary data collection. Rapid rural appraisals or SWOT analysis (a study of strengths, weaknesses, opportunities, and threats facing the sector) may be used in a sector analysis.

Morocco's Institut National de la Recherche Agronomique (INRA) and ISNAR developed the constraint tree's use in agricultural research planning based on the "planning by objectives" approach developed by the German technical assistance organization, GTZ. A visualization technique, known as the "metaplan method," is key in building constraint trees. Basically, the method consists of eliciting ideas from workshop participants, writing these on cards as accurately and synthetically as possible, and posting the cards on a board or wall

Table 1. Examples of Primary Branches in a Constraint Tree

Programs	Technical Constraints			Socioeconomic Constraints	NRM Constraints	Mixed Constraints
	Production	Harvest/Storage	Processing			
Faba bean, Morocco	1. falling yields in Saiss 2. falling yields in Chaouai 3. falling yields in Pré-Rif	4. high harvest and storage losses	5. improper product processing	6. high production costs 7. falling exports		
Olive, Morocco	1. rotation not controlled 2. low productivity in irrigated systems 3. low productivity in intensive systems 4. low productivity in extensive systems		5.deficient processing	6. deficient market structure		
Oil crop, Morocco	1. low productivity in sunflower production in high-potential bour 2. low productivity in rape production in semi-arid soils 3. low productivity in soybean production in irrigated systems		4. deficient processing of annual oil crops	5. lack of organization of the annual crop subsector		
Date palm, Morocco	1. low productivity		2. deficient processing of dates			3. poor use of marginal lands 4. role of date palm in complex production systems
South Benin	1. low productivity			2. poor institutional development	3. excessive use of natural resources	
Central Benin	1. low crop productivity 2. low livestock productivity			2. inadequate development of the subsectors	3. poor management of natural resources	4. poor integration of production systems
North Benin	1. low agricultural productivity				2. degraded natural resources	3. poor land-use management

for discussion. This technique typically improves communication among participants and individuals' capacity to integrate the information tabled.

Identifying the major constraint, expressed negatively (e.g., "declining profitability"), is the first step in the tree-building process. Participants are asked to write on cards what they perceive to be the key constraint to performance of the sector. The moderator collects the cards and clusters comparable ones. The planning group then decides which cluster is most important or combines several clusters into one overarching major constraint.

The next step is to single out the causes of the major constraint, also expressed negatively (e.g., "high production costs"). For every cause identified, all further possible causes are determined, and so on. In this way all cause-and-effect relations that contribute to the major constraint are identified. For large and complex constraint trees workshop participants may form parallel working groups, each concerned with a different primary constraint.

In practice, there is no set rule for determining at what level to end the constraint tree. However, common practice is to end a given branch when the analysis has led to a nonresearchable constraint or to a researchable constraint that is impossible to analyze further, as is the case for the constraint "optimum seeding date unknown." The number of levels may vary according to the branch. The number of levels does not necessarily correspond to the importance of the constraint. Rather it reflects the detail of analysis for a given problem area. In general, a high number of levels reflects the complexity of the problem and the ability of planning group members to be comprehensive in their analysis. For example, constraint trees for Morocco's faba-bean program (figure 1) and for Benin's regional research programs ended at the seventh level. In contrast, the constraint tree for Morocco's date palm research program was carried through to the twelfth level, reflecting the complexity of the problems facing the Moroccan palm industry.

Quality of the constraint tree

A constraint tree contributes most to research program planning if it is relevant and complete. It is relevant when the constraints described in the boxes are explicit, specific, and correspond to needs for new knowledge or technologies. It is complete when the sector is exhaustively analyzed. There are three essential elements for building relevant and complete constraint trees: identification of the major constraint, composition of the planning group, and skill of the workshop moderator.

The major constraint must be specific to the sector under consideration and reflect as accurately and synthetically as possible the overall situation in technical, environmental, and socioeconomic terms. Therefore, the major constraint cannot be vague or too comprehensive, such as "lack of strategy for develop-

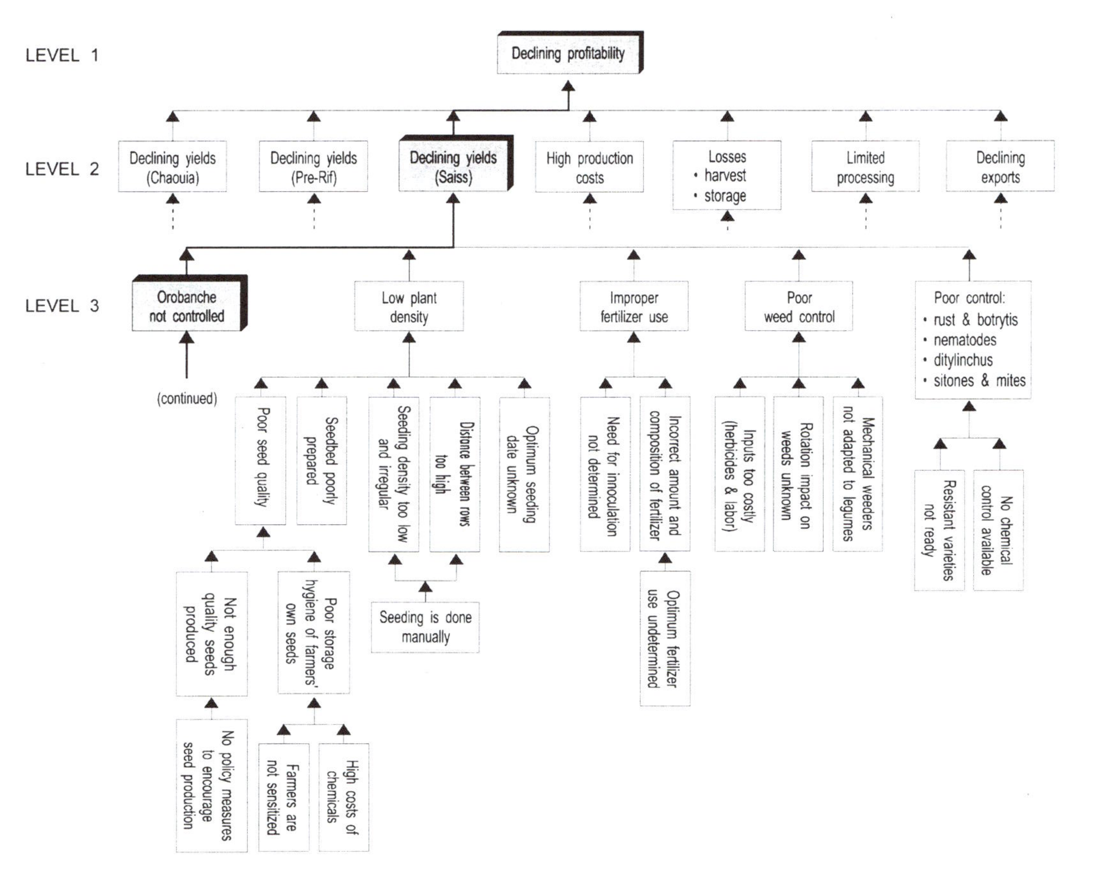

Figure 1. Constraint tree, faba bean sector, Morocco

Note: Orobanche, or broomrape, is a chlorophyll-free parasitic plant living on the roots of other plants, including legumes.

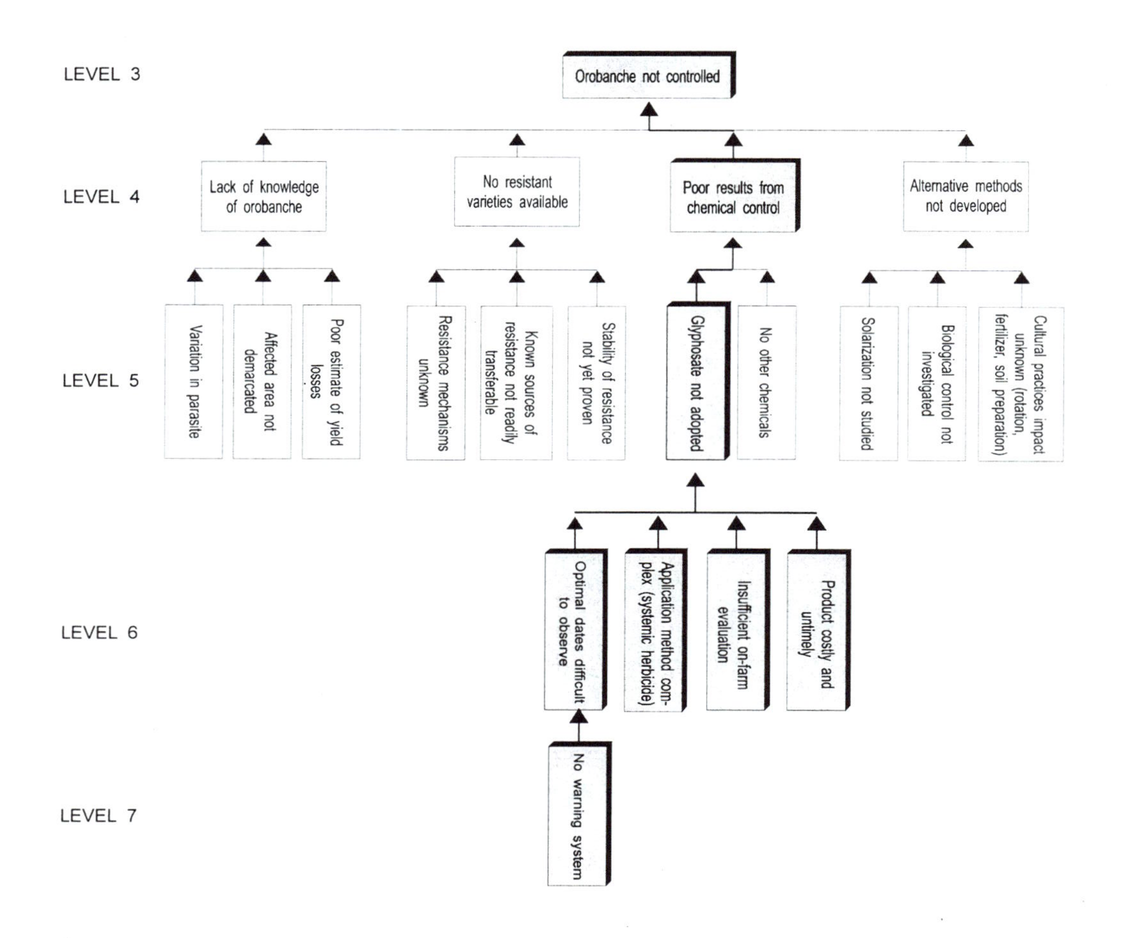

Figure 1. Continued

Note: Orobanche, or broomrape, is a chlorophyll-free parasitic plant living on the roots of other plants, including legumes.

ment of the sector." Neither should it be too restrictive, reflecting purely technical concerns, such as "low cereal yields." For the olive program in Morocco the major constraint might be defined as "the Moroccan olive oil industry faces fierce competition from annual oil crops." This constraint reflects marketing and consumption aspects as well as the technical problems related to the productivity of olive groves and performance of oil mills.

Members of the planning group should be known for their knowledge of the sector and able to listen and articulate their ideas in a heterogeneous group. The planning group should integrate representatives of all the stakeholders concerned with the development of the research program. Normally this includes scientists and research managers, producers, representatives of agribusiness and import-export companies, development workers, and extension workers. All agricultural disciplines should be represented equitably – neither social sciences nor technical disciplines should dominate the process. Broad and complete representation of research disciplines and perspectives reduces the chance that constraints are overlooked or that some aspects are overemphasized. For example, a constraint tree for a small ruminants research program constructed by zootechnicians and veterinarians would probably include only constraints related to breeding and animal health. For the tree to show nutritional constraints, involvement of a forage agronomist might be needed.

Client participation should include all actors in the agricultural sector: farmers, seed producers, nursery managers, food industrialists, traders, and so forth. It is also important to balance the different categories of producers (small, medium, and large scale). Without involvement of small-scale farmers, their problems might be neglected. On the other hand, if medium- and large-scale cereal producers are not represented, other problems may be overlooked, such as those related to pest and disease management and to chemical weeding.

The skill of the moderator is also critical in building a good-quality tree. The moderator should have experience in agricultural research as well as in managing group discussions. He or she should use patience and tact to ensure that all constraints are thoroughly examined and specified. The moderator submits all broad, general constraints, such as "poor performance of plant material" or "inadequate technical processes," for further analysis. Failing this, the constraint tree becomes a "catch-all" type of tree reflecting too many diverse situations. The moderator's experience is particularly put to the test when defining the major constraint and the constraints at level 2 (the primary branches), and towards the end of the process in deciding when to end the analysis. If the planning group is divided into several subgroups and each asked to analyze the constraints related to one or several primary branches, the moderators for the subgroups should be selected based on their knowledge and ability to maintain some distance from their own field of expertise. (Guidelines for organizing a constraint tree-building session are in the ISNAR training module "Research Program Formulation").

Relevance for agricultural research

The use of constraint trees to identify research projects and develop integrated research programs has numerous advantages for research managers: by cooperating in the analysis of constraints facing the sector, a broad-based diagnosis arises from which research plans can be developed. In addition, the method allows for consensus building on the different constraints and for an understanding of the interactions among constraints. In general the constraint tree encourages multidisciplinary thinking and the planning of problem-oriented multidisciplinary research.

Among the constraints identified, there are researchable and nonresearchable ones. Researchable constraints (e.g., "effects of cultural practices such as fertilizer application, soil preparation, and rotation on orobanche are not understood") can be alleviated or removed by application of research results. If a constraint is viewed as researchable, the nature of the constraint may be further defined to decide which organization should do the research or whether the constraint falls within the mandate of the research program involved. Whereas researchable constraints are used to develop research objectives and research projects, the nonresearchable constraints fall outside the scope of the research program. Understanding the nonresearchable constraints is important, however, for two reasons: First, such understanding is useful for agencies collaborating with the research organization and may help them improve their plans. Second, nonresearchable constraints point out the factors that condition the success of the research program. If all the researchable constraints are removed but the nonresearchable constraints still exist, much of the research program's impact will probably be lost. Further, nonresearchable constraints may fall within the competence of development agencies, commercial enterprises, or farmers themselves (e.g., the constraint "lack of early-warning system to control orobanche" or "lack of farmers' associations to help in control of orobanche"). The research program should look for support in other organizations to overcome nonresearchable constraints.

After completion of the constraint tree, the next step is to identify research projects. To do so, the planning group may elaborate a tree of research objectives. The "objective tree" is based on the constraint tree, on perceptions of research opportunities, and on the evaluation of earlier research results (in order to establish realistic objectives that do not duplicate past work). Building a tree of objectives enables planners to visualize, in a diagram similar to that made for the constraints, the results that must be achieved for the various intermediary objectives in order to accomplish the general program goal, which corresponds to the major constraint in the constraint tree.

The constraint-tree approach has proven useful for agricultural research organizations because it can be adapted to different domains: commodity subsectors, groups of commodities, and agroecological zones or regions. It can be applied in the context of a workshop, it is not time consuming, and it is inexpensive in terms of staff requirements (a group of 15 participants is normally able to complete a comprehensive constraint tree in less than three days). Finally, although preparation adds to the quality of analysis, no preliminary work is required of the workshop participants: it's what they have in their head that matters.

There are some limitations to the constraint-tree approach, however. Because it relies principally on judgments made by members of the program-planning group, its results depend largely on the group's composition. Constraints that cannot be directly related to the major constraint are difficult to include. Finally, trees often have too many branches, which may become difficult to manage in later stages and lead to fragmented research programs.

Example

Figure 1 shows the main constraint and primary branches of the constraint tree for the faba-bean sector in Morocco, as well as the specific constraints of primary branch 3: "declining yields in Saiss." This tree consists of seven level 2 branches. Five branches relate to technical constraints (declining yields in Chaouia, Pré-Rif and Saiss; harvest and storage losses, and limited processing) and two relate to economic constraints (high production costs and declining exports). This constraint tree was analyzed to the seventh level.

Level 1 shows the major constraint characterizing the faba-bean sector in Morocco: declining profitability. Level 2 shows the primary causes of the major constraint: falling yields in the three regions where the crop is grown, high production costs, harvest and storage losses, limited processing, and declining exports. Only branch 3 (declining yields in Saiss) is analyzed further here: the presence of orobanche, low plant density, improper fertilizer use, and poor crop maintenance are among the causes of the production decline in this region. These constraints are presented at level 3. At level 4, further analysis of the constraint "orobanche not controlled" reveals that varieties of faba bean resistant to broomrape have not yet been developed and chemical control using glyphosate (commercially known as "Roundup") and other control methods (chemical, biological, cultural practices) have not been adopted.

Pursuing the deficient chemical control, level 5 shows that farmers have made limited use of glyphosate and that research efforts have given little attention to other herbicides. Level 6 shows that chemical control using glyphosate has not been adopted because application dates are difficult to follow (for farmers the appropriate dosage is hard to control and the cost of treatment is high). However, these factors have not been examined thoroughly as there have been insufficient on-farm experiments with this control method. Finally, level 7 shows the glyphosate application date to be difficult to follow, since farmers do not receive information from early-warning systems.

References

Collion, M. -H. and A. Kissi. 1993. Learning by doing: Developing a program planning method in Morocco. *Journal of Public Administration and Development* 13 (3): 261–270.

Collion, M. -H. and A. Kissi. 1994. A participatory, user-friendly method for research program planning and priority setting. *Quarterly Journal of International Agriculture* 33, (1): 78–105.

Collion, M. -H. and A. Kissi. 1995. Guide to Program Planning and Priority Setting. Research Management Guidelines No. 2E. The Hague: International Service for National Agricultural Research.

Division de la Programmation de l'INRA. 1990. Programme de recherche à long terme de l'olivier. Mimeo. Rabat: Institut National de la Recherche Agronomique.

Division de la Programmation de l'INRA du Maroc.1991. Programme de recherche à long terme du palmier dattier. Mimeo. Rabat: Institut National de la Recherche Agronomique.

Division de la Programmation de l'INRA. 1993. Programme de recherche sur les cultures oléagineuses. Mimeo. Rabat: Institut National de la Recherche Agronomique.

GTZ. 1998. ZOPP: An Introduction to the Method. Eschborn: Deutsche Gesellschaft für Technische Zusammenarbeit GmbH.

INRAB. 1996. Plan directeur de la recherche agricole au Bénin. Volume II – Première partie. Plan de développement à long terme. Cotonou, Bénin: Institut National des Recherches Agricoles du Bénin.

ISNAR. 1999. Training Module on Research Program Formulation (day2, session 4, p. 115). The Hague: International Service for National Agricultural Research.

Janssen, W. and A. Kissi. 1997. Planning and Priority Setting for Regional Research: A Practical Approach to Combine Natural Resource Management and Productivity Concerns. Research Management Guidelines No. 4. The Hague: International Service for National Agricultural Research.

Kissi, A. and M. -H. Collion. 1994. La programmation de la recherche par les mots. Mimeo. Rabat: Institut National de la Recherche Agronomique.

Quiro, Z., H. Hilali, C. Kradi, C. Gout, and G. Kleene. 1997: Quelle recherche pour le développement des zones de montagne du Maroc? ICRA/INRAM. Série Document N° 66. Mimeo. Rabat: Institut National de la Recherche Agronomique.

Chapter 23
Tools for Gender Analysis

Gerdien Meijerink, Helen Hambly Odame, and Brigitte M. Holzner

Gender analysis emerged from the recognition that women produce between 60 and 80 percent of the food in most developing countries. The needs of women farmers and relations between men and women are thus relevant in planning agricultural research and development. Gender analysis aims to make social roles and relations explicit. Its purpose is not to create a separate body of knowledge concerning women and agriculture, but to take into account in research and planning other factors and realities that affect, or are influenced by, gender relations. Using gender analysis, it is possible to tailor interventions to meet women's and men's specific gender-based constraints, needs, and opportunities, thereby increasing agricultural research's impact and the effectiveness.

What is gender analysis?

Gender analysis is the study of gender roles and relations in the context of a specific research problem. Whereas "sex" refers to the biological difference between males and females, "gender" implies culturally prescribed roles and identities of men and women. These roles and relations exist between individuals and as characteristics of social structure. Gender is created and changes according to a particular society's assignment of certain activities to women and others to men. Gender analysis also recognizes that roles and relations are cross-cut by other human interaction that implies power relations such as class, caste, age, household position, race, and ethnicity.

Gender analysis addresses women's needs and their participation in the research planning process. It recognizes that women are part of the population for which agricultural research is being planned, and that their representation should be integrated into other analytical tools and processes (e.g., priority setting and constraints analysis). In gender analysis, women cannot be subsumed under unspecific categories such as clients, household, farmers, and user groups.

Gender analysis can be made to fit the scope and scale of various agricultural research plans and planning processes. At the project level, it may explore how agricultural activities, resources, and benefits are distributed at the household level between men and women. At the strategic planning level, it may be used to identify how agricultural research policy impacts gender roles and relations. There is substantial convergence between gender analysis and participatory approaches, and different gender analysis tools are available for use at the different levels of planning. The literature is rife with material about gender analysis tools, including several "how to" manuals as indicated in table 1.

Doing gender analysis

Given the considerable amount of material available on gender analysis training and tools, this chapter limits its focus to two important aspects for planning. First is how to "gender," that is, improve from a gender perspective, existing planning processes and tools such as those presented in this sourcebook. Second is how to create a supportive environment in which gender analysis can be conducted and sustained.

Table 1. Major Gender Analysis Tools for Agricultural Research Planning

Level	Type of Planning	Focus of Analysis	Tools
Research policy (national)	strategic planning	- policy - stakeholder	- content analysis of national development policy and plan statements in reference to women and development (Moser 1993) - policy compared to "prototype" gender policies (FAO 1999B) - sex-disaggregated census data (GENESYS 1996, Hedman et al. 1996) - stakeholders mapping or network diagram (FAO 1999B) - gender analysis framework (March et al. 1999)
Research organization or institute	program planning	- organization - program	- gender analysis matrix (CEDPA 1996) - gender constraints analysis (Moser 1993, FAO 1999B) - gender cost-benefit analysis (Kabeer 1997) - gendered logical framework (CEC 1993, Kerstan 1995, ISNAR 1996) - gender analysis activity profile (ECOGEN 1993, FAO 1999B)
Research project (multi-year)	project planning	- community - household	- practical and strategic needs assessment (Moser 1993) - seasonal calendar of gender-based activities (Feldstein and Poats 1989, Feldstein and Jiggins 1994) - participatory gender resource mapping (Willmer and Ketzis 1998, FAO 1999B) - livelihood analysis (FAO1999)
Research project (annual)	annual planning and budgeting	- benefits	- application of gender analysis to on-farm trials (Feldstein and Poats 1989) - benefits analysis flow chart (ECOGEN 1993)

Tools for "gendering" planning tools and processes

While some tools for gender analysis are unique, others are based on existing analytical tools that are "gendered" or improved by incorporating gender analysis.

Gendered logical framework. The gendered logical framework makes explicit the gender impacts of a program or project. Here, women stakeholders are involved in developing the logical framework matrix. In the logical framework's six basic steps (see Baur, this volume), the following general questions can be added:

- Do the goal, purpose, objectives; outputs and inputs; or activities address the problems of women and men? Do specific inputs or outputs implicate gender relations? How accessible is the planned program to women and how appropriate is the project to the strategic needs of women?
- Do the critical assumptions imply greater risks for women or men?
- Can quantitative and qualitative sex-disaggregated or gender-responsive indicators be defined for monitoring and evaluating the program or project (e.g., number of women participants in the activity, extent of women's participation, degree of women's influence in decision making in the activity)?

Questions for gendering each level of the logical framework are detailed in CEC (1993: 22–23) and ISNAR (1996).

Gendered constraint analysis. Constraint analysis usually begins with formulation of a central problem or major constraint. In the ensuing steps, the causes of the major constraint are identified, which themselves are broken down into their main causes (see Kissi, this volume). Incorporating gender analysis into constraint analysis entails identifying the constraints related to gender. Three questions can guide this analysis:

- What is the major constraint and does this problem implicate gender roles or relations?
- What are the gender-related causes for this constraint?
- What are the gender-related consequences of the constraints?

For example, if the major constraint is "low cash-crop productivity," then possible gender constraints could be "men do mostly off-farm work," "women have no access to revenue from cash crops," "women face time constraints," or "women have difficulty accessing credit."

Gendered geographic information systems. A geographic information system (GIS) is a tool for managing and visualizing large amounts of (socioeconomic) data. As such, it helps decision makers analyze and prioritize research problems (see Pachico, this volume). GIS is useful for storage and presentation of sex-disaggregated data and gender statistics. A gendered GIS might show regions where the number of female-headed households is particularly high or low; or it could show walking distance for collecting firewood or water. Such data helps planners identify gender issues in agricultural problems and needs for improvements in conditions facing rural women and men (Hedman et al. 1996). Here too a number of questions can guide the "gendering" process:

- What relevant gender-specific regional data should be assembled (e.g., poverty incidence associated with female-headed households, land rights, male or female migration, feminization of agriculture, literacy of women and men)?
- What relevant gender-specific, village-level data should be assembled (collecting points for firewood or water, markets, areas where "women's" and "men's" crops are grown)?
- How have gender-relevant variables changed over time in the regions being analyzed (e.g., changes in literacy rates for women and men)? How have conditions for women improved? What gender constraints are evident?

Gendered cost-benefit analysis. Cost-benefit analysis is a planning tool that can be used at the level of policy interventions, at the research program level, or at the project level. Cost-benefit analysis weighs the costs and the benefits through time using a discount factor. Costs and benefits that accrue at a later stage in time are valued less than more immediate costs and benefits. These weighted or "discounted" costs and benefits form the net present value (NPV) of a policy, program, or project.

Costs and benefits are not necessarily equally or fairly distributed across a society. Men and women may accrue different kinds of benefits. Similarly, a benefit for a male farmer may be a cost for a female farmer or vice versa. In determining opportunity costs, it must be remembered that opportunity costs (e.g., cash or labor costs) may differ for women and men and that all costs, tangible and intangible, must be determined. Women workers, for example, may not be paid wages. But this does not necessarily mean that harnessing their work implies no opportunity cost. Intangible benefits, such as reduced workload, are typically of great value to women. However, a project may unintentionally increase the labor burden of one group of women and not another. Overall, the gender division of resources and benefits is usually asymmetrical within and outside of the household. Cost-benefit analysis must therefore build this consideration into data collection and analysis.

The gendered cost-benefit analysis identifies costs and benefits that have gender implications. A number of straightforward questions can guide the analysis:

- Which benefits accrue to men and which to women? (It is important to keep track of all benefits associated with a project.)

- Which costs are borne by men and which by women?
- What are the opportunity costs to men and to women?

Despite efforts to "gender" the cost-benefit analysis, considerable difficulties remain in recognizing and working with nonmonetary and nonquantifiable costs and benefits. Kabeer (1997: 163–186) has argued that these considerations are critical in the planning cycle. In this respect, cost-benefit analysis should be approached with caution, and is best used in combination with planning tools, such as participatory appraisals, that can recognize nonmonetary, qualitative considerations in the planning cycle.

Gendered participatory appraisals. Participatory approaches are not necessarily cognizant of gender relations (Mayoux 1995). Yet in participatory rural appraisal (PRA), gender must be made explicit. Tools for participatory appraisal (see Henman and Chambers, this volume) such as seasonal calendars, Venn diagramming, and matrix scoring are useful for illuminating gender issues for agricultural research planning. They are also often appropriate for identifying and working with qualitative data.

PRA tools can be used to learn who does what, who owns what, and who has rights to what. An example of a PRA tool is the development of a seasonal agricultural calendar with both men and women farmers. Here, men and women are first separated and each group develops its own labor calendar. The two calendars are then presented side by side and discussed by women and men together. This exercise helps groups identify division of activities and differences in time allocations between men and women. It often points out both complementary and conflicting activities in the schedule (including labor and capital requirements). This type of tool often illustrates women's triple work burden, as they are responsible for household, farm, and community development activities, including agricultural marketing (Feldstein and Jiggins 1994).

Another PRA tool, Venn diagramming, is used to explore the different networks and connections between certain problems and male and female stakeholders. Matrix scoring, which is used to examine people's own criteria for choosing among options can reveal how men and women use different criteria in making decisions and, therefore, may assign different priorities to a technology or natural resource (e.g., a new seed variety or soil conservation technique).

Creating an environment for sustaining gender analysis

Incorporation of gender analysis in any level of agricultural research planning will be limited without simultaneous efforts to create a research environment that is responsive to gender analysis. Creating such an enabling environment in agricultural research planning is a process and it involves different layers of organizational change. Table 2 summarizes essential components and activities in the change process.

Awareness is the foremost critical component in creating an environment conducive to gender analysis. Without awareness, agricultural research topics or agricultural technologies appear gender neutral. Discussions on gender issues should, therefore, focus on actions to improve gender-related problems. Leadership of top managers and scientists is a second crucial component for gaining and sustaining support for gender analysis. Rather than appointing a solitary "gender specialist" in the organization, sustained capacity for gender analysis is best achieved by "mainstreaming" gender in the organization's policies and processes.

Ongoing data collection and analysis is also vital, and staff should be made responsible for this task and given the resources to monitor data collection. Findings of gender analysis should flow directly into all planning processes and become an integral part of the organization's work. Systematic monitoring and evaluation of gender-related impacts complement ongoing data collection. Reports on gender analysis document results and widen the support base for future projects and funding. The importance of gender issues needs to be assessed in conjunction with other concerns such as subsistence and commercial farming and different regional needs.

Table 2. Components and Activities for Creating An Enabling Environment for Gender Analysis

Critical Components	Activities
Awareness	– conduct gender awareness training – collect and distribute reference materials on gender issues and tools
Commitment to action that addresses gender issues	– gain support of top managers
Capacity to formulate research hypotheses that are gender responsive	– conduct gender analysis training – link with national and international networks involved in gender-related work
Capacity for carrying out gender analysis	– designate human and financial resources for gender analysis
Application of findings of gender analysis to the organization's planning processes	– integrate gender analysis into all levels of planning
Systematic monitoring and evaluation of gender impact	– plan for the regular evaluation of gender-related impact
Regular reporting of research results and "lessons learned"	– produce research reports and press releases on the findings of gender analysis

Relevance for agricultural research

Gender analysis improves the effectiveness of agricultural research. Gender analysis emerged from the recognition that women produce a large proportion of food in most developing countries (FAO 1999b). In some regions, women form the majority in rural populations and households. However, women are usually socially, politically, and economically disadvantaged relative to men. Research organizations have seen many projects and activities fail because gender roles and relations were not taken into consideration at one or all levels of planning. Moreover, agricultural research and extension have often failed to improve the conditions and positions of rural women.

The efficiency of agricultural research can also be improved through gender analysis. Agricultural technologies are seldom as gender-neutral as they may seem. By default, they are often designed with and for male farmers and may depend on resources that are not easily accessed and controlled by female members of a household or community. It has been recognized for more than 25 years that the neglect of gender issues in agricultural research can lead to situations whereby technology is left unused or used inappropriately (Staudt 1975). Agricultural research may inadvertently create new problems when it involves men in agricultural production activities that were formerly the domain of women (Agarwal 1992). In this way, female farmers may lose access to resources (Mbilinyi 1992). Neglect of gender in agricultural research is thus an inefficient use of human and technical resources, as well as being detrimental to women's material well-being and economic and political status in society (Young 1988).

Examples

Kenya

At the Kenya Agricultural Research Institute (KARI), gender analysis has been used to bring to the forefront of research planning, implementation, and evaluation the relations of men and women farmers and their different knowledge, perspectives, and needs. KARI realized in 1995 that efforts to institutionalize gender awareness and analysis needed to be harmonized. That same year it established a gender task force. This group consists of representatives of all KARI's research departments and its major supporting organizations and donor programs. In 1996, the task force produced its first three-year action plan with the aim of incorporating a gender perspective in all phases of research. The gender task force has become a strong working committee in the institute for several reasons:

- Gender analysis is integrated into the farming systems approach to research, extension, and training emphasized by KARI.

- Top management, almost all of whom were men, established the gender task force in 1995 and have maintained their involvement by providing guidance and resources for its activities.
- Training is used to create gender awareness throughout KARI's 12 research centers.
- To build capacity in gender analysis within the institute, KARI gender trainers were trained to the degree level; external specialists in gender were also called upon for inputs, but the process was led by institute, through its own gender task force.
- KARI created its own set of gender guidelines for project management.
- Conferences, information sharing, and publishing relevant research results have spotlighted KARI's achievements in the area of gender analysis.

Work done by the gender task force reveals the organization's efforts to sustain an environment supportive of gender analysis. Simultaneously, KARI has worked to strengthen its capacity in terms of trained trainers, to introduce gender analysis tools into research planning, and to further adapt such tools to the organization's needs.

Costa Rica

In 1996, the Costa Rican government undertook to strengthen the productive role of rural women and to improve their livelihoods. The agriculture and livestock ministry and government offices related to women's affairs collaborated to introduce and develop a gender approach in the policies, programs, and activities of the mixed farming and environmental sectors. A gender planning committee implemented the project.

Sensitizing, motivating, and training technical and administrative personnel and farmers was a major component of the initiative. Besides training, grassroot groups and institutional representation were strengthened to support adoption of a gender perspective at all levels. Policies were revised by first identifying problems, then indicating actions, measures, and institutional mechanisms to use to solve the problems and amending policies that had differential impacts on men and women.

This initiative fostered a climate of cooperation and exchange between officials and farmers (male and female). The project also resulted in the appropriation of gender tools by agrarian policymakers and technical personnel linked to rural development projects. The project therefore did not remain an isolated one-off activity. Another major result was that indicators were specified for systematic and coherent measurement of progress made in gender issues.

Nepal

In 1996, the women farmers' development division of Nepal's ministry for agriculture embarked on a pilot project to improve information on women's contributions in agricultural production. Its wider goal was to make planning gender sensitive. Impetus for the project was the perception that agricultural planners and extension personnel often neglected rural women's needs as producers. The project, therefore, aimed to improve information on rural women and men and to involve them in local planning in the agricultural sector.

Several PRA case studies were conducted in each agroecological zone on gender issues in Nepali farming systems. Of these PRAs, three short video films were made. Using these films, district-level staff were trained in gender sensitive participatory approaches to planning. Guidelines were also formulated for gender-responsive agricultural planning at the district level. The project was completed with a national-level workshop at which policymakers and ministry staff discussed the guidelines and encouraged their adoption into the ministry's planning processes.

References

Agarwal, B. 1992. The gender and environment debate: Lessons from India. *Feminist Studies* 18 (1): 119–158.

CEC (Commonwealth Economic Comission). 1993. Project Management. London: CEC.

CEDPA. 1996. Training manuals in gender and development. http://www.cedpa.org.

ECOGEN (Ecology Community Organization). 1993. Training Manuals. Virginia: Clark University and Virginia Polytechnic Institute and State University.

FAO. 1999a. Gender and food security: Agriculture. Online. http://www.fao.org/gender/en/agri-e.htm.

FAO. 1999b. SEAGA Field Handbook Socioeconomic and Gender Analysis Programme. Rome: Food and Agriculture Organization of the United Nations. (See also the SEAGA intermediate and macro handbooks for planners and policymakers, http://www.fao.org/sd/seaga/SEfh0001.htm).

Feldstein, H. S. and J. Jiggins. 1994. Tools for the Field. New York: Kumarian Press.

Feldstein, H. S. and S. V. Poats (eds). 1989. Working Together: Gender Analysis in Agriculture. Vols. I and II. West Hartford: Kumarian Press.

GENESYS (Gender in Economic and Social Systems Project). 1996. Gender Analysis Tool Kit. Washington, D.C.: US Agency for International Development's Office of Women in Development.

Hedman, B., F. Perucci, and P. Sundström. 1996. Engendered Statistics: A Tool for Change. Stockholm, Sweden: Statistika centralbyrån.

ISNAR. 1996. SADC/ESAMI/ISNAR Training Module on Gender Analysis in Research Management of Agriculture and Natural Resources. The Hague: International Service for National Agricultural Research (See also http://www.cgiar.org./isnar/activities/training/selection.htm)

Kabeer, N. 1997. Reversed realities: Gender hierarchy in development thought. Editorial. *IDS Bulletin*, 23 (3): 163–186.

Kerstan, B. 1995. Gender-sensitive Participatory Approaches in Technical Co-operation. Eschborn, Germany: Deutsche Gesellschaft für Technische Zusammenarbeit.

March, C., I. Smyth, and M. Mukhopadhyay. 1999. A Guide to Gender Analysis Frameworks. London: Oxfam.

Mayoux, L. 1995. Beyond naivety: Women, gender inequality and participatory development. *Development and Change*, 26 (2): 235–258.

Mbilinyi, M. 1992. Women and the pursuit of local initiatives in Eastern and Southern Africa In *Reviving Self-reliance: People's Response to the Economic Crisis in Eastern and Southern Africa* edited by W. Gooneratne and M. Mbilinyi. Geneva: United Nations Centre for Regional Development.

Moser, C. 1989 Gender planning in the Third World: Meeting practical and strategic gender needs. *World Development,* 17 (11): 1799–1825.

Moser, C. 1993. Gender Planning and Development: Theory, Practice and Training. London: Routledge.

Staudt, K. 1975. Women farmers and inequalities in agricultural services. *Rural Africana*, 29: 81–94.

Willmer, A. and J. Ketzis, 1998. Participatory Resource Mapping: A Case Study in a Rural Community in Honduras. PLA Notes No. 33. London: International Institute for Environment and Development.

Young, K. 1988. Gender and Development: A Relational Approach. Sussex: Institute of Development Studies.

Chapter 24
Geographic Information Systems

Douglas Pachico

Geographic information systems (GIS) help planning participants to manage and visualize large amounts of data. Biophysical and socioeconomic data, spatially referenced, are stored in a database. The database is then used to draw maps reflecting the distribution of production, marketing, and environmental or socioeconomic problems. Such maps enable decision makers to analyze and prioritize research problems. GIS can also be used to define homogenous regions in which research efforts can be effectively targeted and to identify regional similarities and the potential for technology transfer. An easily accessible GIS improves consistency in planning across the different planning levels and research programs. GIS is a tool generally used in conjunction with other planning tools, such as simulation and economic modeling. It can be incorporated in a large array of planning approaches.

What is geographic information system analysis?

Geographic information system (GIS) analysis makes use of spatially referenced data that can generally be displayed as map images. A GIS can be used in agricultural research planning for a range of purposes, such as prioritizing production constraints, selecting research sites, and targeting of research results to a wider area.

While the term GIS has been in use since the 1960s (Bracken and Webster 1990), common usage of GIS in agricultural research planning is considerably more recent. Routine use of GIS has become practical only in the last decade or so due to advances in computerized data processing, data storage, and software. Advances in remote sensing, by which images of the Earth are obtained from satellites, have also expanded horizons for GIS, greatly adding to the data available at reasonable cost. Continued technical change in these fields can be expected to make GIS ever more affordable and thus more accessible to a wider range of partners in planning agricultural research.

GIS databases can be designed at different levels of aggregation, from the continental and countrywide levels to regional, microwatershed, or even village levels. Data from different scales can be nested hierarchically to relate analyses at different levels of aggregation. Satellites have been instrumental in improving geographical positioning technology and have expanded opportunities for obtaining microlevel field data. This, in turn, can be related to more general, higher level regional or national GIS data.

A wide variety of types of data relevant to agricultural research planning can be georeferenced, such as temperature, rainfall, soils, land use, crop distribution, water courses, official boundaries, transportation networks, population, and human welfare indices (e.g., heath, education, income).

GIS is more a tool for handling and representing large amounts of complex data than an analytic method grounded in a systematic theory or causal understanding of phenomena. Although GIS itself may not always generate hypotheses, it can serve as an excellent cross-disciplinary tool that agronomists, geneticists, soil scientists, physiologists, economists, and social scientists can use to analyze data in order to produce insights valuable for agricultural research planning purposes.

The potential power of GIS analysis thus lies in its capacity to spatially relate a range of different variables and to do so at different scales. Despite the great underlying complexity of GIS in both data and analysis, much of the resulting information can be communicated intuitively to nonprofessionals as colored maps, making it an especially attractive tool for use both by senior decision makers and in interdisciplinary planning events.

GIS has been used to address a number of research planning issues, notably the following:

- identifying where a specific technological innovation is likely to be relevant by defining relatively homogenous agroecozones
- appraising the importance of a constraint by estimating the area over which it prevails
- improving the targeting of technology development through enhanced understanding of the conditions that characterize the places where target groups are concentrated (e.g., poor farmers, woman-headed households)
- selecting research sites, either on station or on farm, that best represent some combination of conditions (e.g., rainfall, soils, crop system)
- assessing the degree of confidence with which results of research at one site can be extrapolated to other sites with varying degrees of similarity to the research site
- estimating expected returns to research and portraying their spatial distribution through linkages to economic and simulation models

Clearly GIS can be an extremely useful tool in making research planning decisions. However, its utility lies in its use in conjunction with other approaches. Alone it is neither comprehensive nor self-contained as a decision or planning tool.

Using geographic information systems

Agricultural research planning typically utilizes GIS as one stage in a process that involves other techniques and approaches as well. GIS is generally used early on in the planning process, for example, to identify target regions. When GIS includes data on crop distribution by region, it can be linked to economic models that permit estimation of the value of outputs of the agricultural research system. Linking models of expected impact of research to other GIS data yields an image of the distribution of benefits that can be anticipated from a new technology.

The effective use of GIS requires an initial investment in building a georeferenced database. Especially at the national level, this can often be achieved by using existing secondary data complemented by remote sensing images. The effort of compiling and organizing a georeferenced data set can still be substantial, however, and available secondary data are unlikely to be without shortcomings or gaps. Ever more sophisticated and agile methods are being developed to compile georeferenced databases efficiently and to exploit more completely the information already available.

Decisions about the degree of detail needed in the data and the extent to which particular variables must be included in the GIS require careful planning and explicit, clear choices. In a sense, there are always reasons to incorporate additional data, and without doubt the richer the database, the greater the potential value of the GIS. Yet additional information implies costs of acquisition, entry, and updating, so inevitably priorities must be set and data inclusion limited in the GIS.

Even though the agricultural research system can extract great value from GIS databases, often it will be more appropriate for other agencies to share responsibility for developing different components of the system. For example, ministries of agriculture typically conduct agricultural censuses; ministries of environment may collect data on soils, water, and biodiversity; ministries of health may collect food consumption data; and specialized census bureaus are often responsible for population data. All this data may be useful for agricultural research planning. Yet in many cases it is beyond the scope of the agricultural research system to pull together all this data into an integrated GIS from which it could derive benefit. A coordinated, interinstitutional, national GIS is perhaps the best approach.

Considerations mentioned thus far relate to a national GIS. But the approach is also relevant at both a higher, transnational scale and at more disaggregated regional and local scales. As public resources for agricultural research have generally become scarcer and the importance of new, sophisticated methods like biotechnology has grown, countries have increasingly looked to regional research networks like ASARECA and SACCAR in Africa and the PROCIs or

FONTAGRO in Latin America to share the costs and benefits of research by better capturing spillovers and achieving economies of scale through division of labor.

GIS can be useful for identifying opportunities for transnational research cooperation. Assessment of the extent to which countries share a specific constraint, identification of homogenous noncontiguous agroecological zones, selection of research sites, and extrapolation of results are all critical elements in developing and implementing a transnational agricultural research agenda. Development of a subcontinental or continental GIS can thus be a high-return investment. Such an effort is best based on a viable national GIS, but it also requires an additional international effort to harmonize national databases.

At a subnational or regional scale, GIS can again be a useful tool for research planning, particularly in large or highly diverse countries. To understand or monitor changes in natural resources, like soils or water within a watershed, GIS can greatly enhance the usefulness of simulation models for appraising the potential impact of alternative technologies or policies. Likewise, GIS on a more local scale provides insights about diversity in soils, climate, or cropping systems that are masked in higher scale analysis, for example, at the national level.

It might seem plausible to assume that the information that can be gleaned from GIS declines as it is applied to ever smaller, more localized (and, therefore, presumably more homogenous) areas. However, GIS is now being applied at the field level in some highly technified agricultural systems where, for example, fertilizer doses are adjusted to conditions in different parts of the same field. GIS across scales is thus highly fractal, with additional diversity, which can be highly significant, observable in detailed analysis.

To be optimally effective for research planning purposes, GIS analysis probably needs to be cross scale. For example, the value of data at a very local scale can be greatly enhanced if it is linked to a higher scale analysis that permits extrapolation of findings from the localized site to a wider area. At the same time, the cost of collecting and managing data on a large scale becomes prohibitive over ever wider zones. Thus, to be cost effective for planning purposes, GIS should comprise a judicious combination of small amounts of data covering a wide area with more data in greater detail covering a few selected small areas.

A particular strength of GIS lies in its capacity to overlay a variety of data. Thus, to define regions within a country, GIS might combine agroecological, socioeconomic, and administrative data (Janssen and Kissi 1997). While GIS applications in agricultural research planning are often dominated by climate or soils data rather than socioeconomic or administrative data, the latter are often crucial. For example, administrative or political boundaries are important in defining target zones because they form the framework through which public services like extension and credit are implemented.

In its initial stages of development and application, GIS for agricultural research planning is likely to make particular use of climate and physical data, like soils, elevation, and water courses. To some extent this is because the characterization of agroenvironments is one of GIS's most useful applications. Also, this type of georeferenced data is relatively widely available. In addition, climate and geology change relatively slowly over time, largely alleviating the need to update this type of data. For most practical purposes, once this type of data has been incorporated into a GIS, the job is done.

In contrast, data for variables that are more a product of human activity, from land use and crop distribution, to transport infrastructure, population, prices, and malnutrition are both less readily available in georeferenced form and almost certainly bring greater maintenance requirements. Without current information on such variables, GIS is less reliable as a research planning tool. Consequently, for the usefulness of GIS to remain optimal over time, some system of updating key variables is essential.

Relevance for agricultural research

GIS can be a helpful tool for national systems in their agricultural research planning. Because it relies on systematic quantitative data, it provides a more objective basis for comparing alternatives than informal judgments. GIS has the capacity to overlay a complex variety of ecological, socioeconomic, and administrative data and communicate results intuitively using maps that are easily accessible to nontechnical planning participants.

Although some specialized expertise is needed to assemble and use GIS, advances in software are making it increasingly accessible to an array of potential users. Similarly, though GIS depends heavily on electronic data capacity, technical advances are lowering computing costs and thereby the threshold for use of GIS by agricultural research organizations.

While constraints related to human capital and computing power are likely to recede in the future, significant initial investments to develop an extensive database are nonetheless needed to exploit the approach. Judicious decisions must be made as to the requirements of the GIS database, taking into account the need to update the system periodically.

A well designed GIS will have the flexibility to address a wide variety of planning questions while providing consistency in analysis over time. A systematic and comprehensive approach to the design and construction of a GIS is likely to be far more useful and less expensive in the long run than sporadic use of GIS on a case-by-case basis. There may be scope at the national level for interinstitutional collaboration and cost sharing in development of a GIS, because agricultural research organizations will seldom be the only or even the main users of the system.

The initial attraction of GIS is likely to be for research planning at the national level. Consequently most NARS will be inclined, and rightly so, to focus first on development of a national scale GIS. The full power of GIS, however, is best attained by linking a national system to systems at other scales.

Even in large-scale systems, such as a national-level one, detailed data from a modest number of individual sites greatly enriches the GIS. Although gathering detailed data on complex processes, such as changes in soil quality, biodiversity, or farming systems, is generally far too costly, a well designed GIS can use such data collected at a few sites to extrapolate general trends.

Likewise, to capture spillovers from research conducted elsewhere, GIS can be usefully shared among countries at a transnational level, to appraise common constraints and design cooperative international research programs. For all but the very largest national agricultural research organizations, transnational cooperation in research is likely to be far more cost effective and feasible than attempts to be self-sufficient nationally in agricultural research (see also Perrault, this volume). GIS can play a vital role in supporting development of a cooperative research agenda.

GIS is generally used in a national planning context in conjunction with other planning tools, including simulation and economic modeling. While it is a useful adjunct to other approaches, in itself GIS is not a stand alone comprehensive planning methodology. Because GIS can be used with a wide variety of different data, linked systematically to other planning tools, and communicated fairly easily to nonspecialists, it can be incorporated into a number of different approaches to planning. Further, it can be tailored to the specific needs and interests of different countries while permitting broad participation in the planning process.

Examples

The most common use of GIS in agricultural research planning has probably been in the identification of target zones. Classically this is rooted in an initial appraisal of the relationships between crop performance and temperature, rainfall, solar radiation, and soils (FAO 1978). Based on this information, agroecological zones are defined and mapped. Agroecological zones are fixed and all-encompassing in this classic approach. Each geographic location belongs to a specific agroecological zone depending on climate and soil characteristics, and this generic zonification holds for all crops and production systems at that location. The example in figure 1 illustrates this approach in identifying target zones for water efficiency research. The case study elaborated in the figure shows how to use GIS to assess the future importance of water-saving technologies for a certain area. It shows how multiple variables such as soil textures, con-

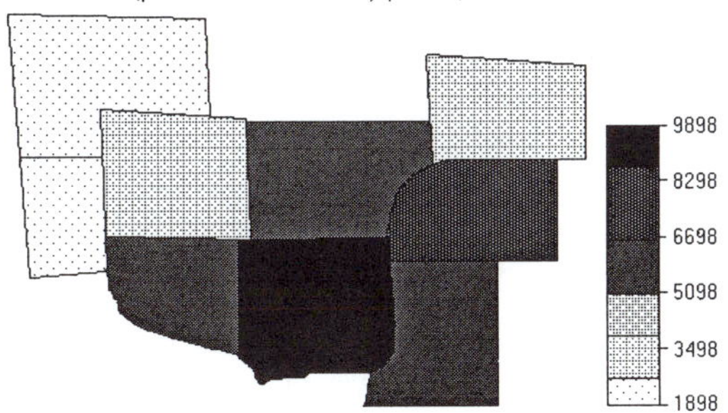

Evaluation of percolation loss (I). The research target area consists of a number of irrigation districts (blocks). First, these irrigation districts are *crossed* with the soil texture map to determine the area of each soil texture class in each district. Percolation losses differ per soil texture class and are shown per irrigation district in 10^{-2} mm.

Conveyance loss (per block for a 10-day period)

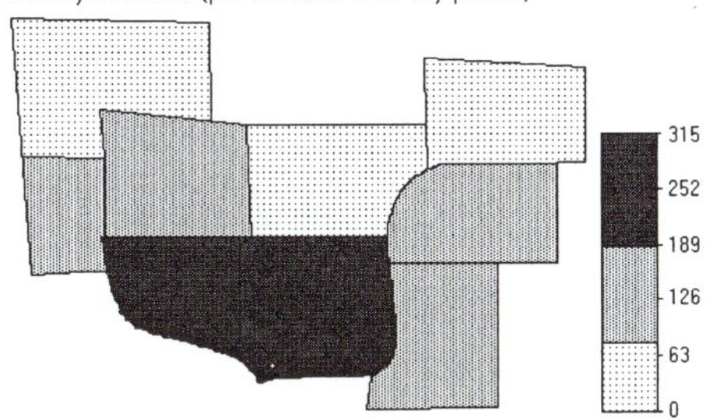

Evaluation of conveyance loss (S). Conveyance losses are calculated in about the same way as the percolation losses. First, the map with the irrigation districts is crossed with the channel distribution map. The conveyance loss per meter channel length differs per channel type and is 0.2 m^3 per day for clay channels and 0.01 m^3 per day for concrete channels. Losses are shown in 10^{-2} mm.

Maximum Evapotranspiration (per block for a 10-day period)

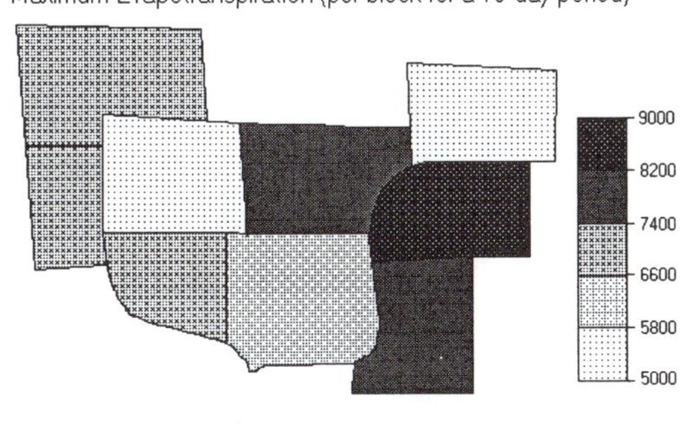

Evaluation of maximum evapotranspiration (ETm). Crop water requirements are normally expressed by the rate of evapotranspiration (ET). The evaporative demand is the reference crop evapotranspiration (ET_0), which predicts the effect of climate on the level of crop evapotranspiration. Here the ET_0 is 8 mm/day. Maximum evapotranspiration is when water is adequate for unrestricted growth and development under optimum agronomic and irrigation management. Evapotranspiration is shown in 10^{-2} mm.

Total water requirements (per block for a 10-day period)

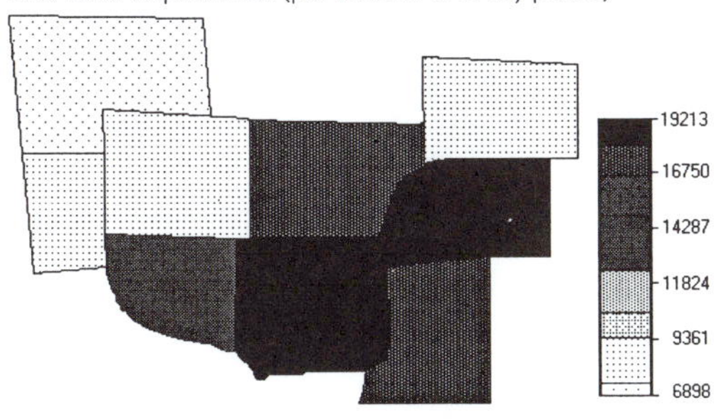

Water requirements calculation (S+I+ETm). The water requirements (in mm) for each of the districts is equal to the *sum* of water losses due to infiltration through the soil (I), seepage through the channel (S) and maximum evapotranspiration (ETm) for each district. For the crop concerned the water requirements for a 10-day period are at or above the future available irrigation gift (75 mm) in 8 of 9 districts. Research to economize water use thus appears to be very promising.

Figure 1. Example of use of GIS to assess the relevance of water efficiency research

veyance loss conditions, evapotranspiration and irrigation gifts are combined to calculate a simple water balance.

An alternative, more flexible approach to agroecological characterization is to use GIS data and specific quantitative criteria to delineate particular, for example, commodity-specific, agroecological zones (Wood and Pardey 1998). The specific thermal or moisture levels that critically affect the physiology and productivity of different crops (e.g., rice and maize) or the prevalence of particular insects or diseases, are unlikely to correspond universally to the parameters defining generic agroecozones. Thus, two locations might fall within the same ecological classification for one crop, constraint, or technology, but in different classes for another. This approach was used by the Kenya Agricultural Research Institute (KARI) (Kamau et al. 1997). KARI established research target zones for eight commodities based on key environmental determinants of productivity of each commodity. This flexible approach to the delineation of agroecozones corresponding to the characteristics of a given crop or production constraint is likely to be most useful at the research project planning level. Nonetheless, at a more general or aggregate strategic planning level, senior decision makers may find generic zonification an attractive simplification.

Agroecozones can be defined by a set of threshold levels in critical variables, for example, values for rainfall or temperature (FAO 1978). This approach may be adequate for most purposes, particularly for individual commodities or constraints. However, if the objective is to describe relative similarities between complex agroecological systems, more sophisticated statistical techniques have to be utilized (Jones 1993). Various statistical approaches such as factor analysis or canonical correlation analysis can be used to characterize zones based on the degree of similarity in a large number of variables.

Even without sophisticated statistical analysis, GIS can overlay a wide variety of data, leading to the definition of thousands of zones that are homogenous with respect to the variables used. This complexity can be reduced by pinpointing which variables really make a critical difference with respect to the matter under consideration, be it sorghum productivity or soil erosion. Expert judgments to determine the key threshold levels in the critical determining variables are essential in this regard (Janssen and Kissi 1997).

Regardless of whether the initial zonification is based on multi-variate statistical analysis or on simple threshold levels of key variables, the preliminary zonification is typically tested against expert judgments of knowledgeable scientists, extensionists, and research managers. Their participation in defining what is a useful zonification system not only supplements a purely mechanical analysis (which can be overly complex or misleading), but also contributes to winning acceptance of GIS-based zonification as a useful agricultural research planning tool.

Definition of agroecological zones can be done at different scales or levels of resolution, each serving different planning purposes (Hunt 1993). For example, to plan agricultural research at the international level of the Consultative Group on International Agricultural Research (CGIAR), nine agroecological zones were defined along lines corresponding to political boundaries based on two thermal classes and four length of growing season classes. While such broad demarcation may be useful for allocating resources among commodities or broad ecozones at a global level, to plan research on a single crop a more detailed analysis is more appropriate. For example, for rice, 15 temperature zones and 21 length of growing season classes were combined to create a matrix of 315 cells. These, in turn, were overlaid with data on four rice production systems. This approach facilitated an accurate understanding of the spatial implications of alternative rice research strategies than would be possible using the broader global analysis.

References

Bracken, I. and C. Webster. 1990. Information Technology in Geography and Planning: Including Principles of GIS. London: Routledge.

Gonzalez, J., B. Gutierrez, P. Jativa, H. Medina, R. Pacheco, and S. Wood. 1998. Evaluación ecológica-económica regional de prioridades nacionales de investigació agropecuaria de los países andinos. Washington, D.C.: International Food Policy Research Institute.

FAO. 1978. Report on the Agroecological Zones Project. Rome: Food and Agriculture Organization of the United Nations.

Hunt, E. D. 1993. Agro-ecological zones for planning and research: Using geographic information systems. *Quarterly Journal of International Agriculture,* 32 (4): 388–408.

Janssen W. and A. Kissi. 1997. Planning and Priority Setting for Regional Research: A Practical Approach to Combine Natural Resource Management and Productivity Concerns. Research Management Guidelines No. 4. The Hague: International Service for National Agricultural Research.

Jones, J. W. 1993. Decision support systems for agricultural development. In *Systems Approaches for Agricultural Development.* Proceedings of the international symposium, Bangkok, Thailand, December 1991, edited by F. Penning de Vries, P. Teng, and K. Metselaar, pp. 459–471. Dordrecht: Kluwer Academic Publishers.

Jones, P. G., D. M. Robinson, and S. E. Carter. 1990. A GIS Approach to Identifying Research Problems and Opportunities in Natural Resource Management. Cali, Colombia: International Center for Tropical Agriculture.

Kamau, M., D. Kilambya, and B. F. Mills. 1997. Commodity Program Priority Setting: The Experience of the Kenya Agricultural Research Institute. The Hague: International Service for National Agricultural Research.

Wood, S. R. and P. G. Pardey. 1998. Agricultural aspects of evaluating agricultural R&D. *Agricultural Systems,* 57 (1): 13–41.

Chapter 25
The Logical Framework

Henning Baur

The logical framework is a tool for planning programs and projects in the broader context of development goals. By leading planners step by step through cause-and-effect relationships from activities to outputs, to goals, it links program inputs and objectives in a clear, logical way. Research managers can use the logical framework to connect research program and project objectives to national goals. A logical framework captures and documents the collective thinking behind a project and guides subsequent investments, monitoring, and evaluation.

What is a logical framework?

Logical framework analysis is a means of planning, implementing, and evaluating research programs and projects. A logical framework (or "logframe") consists of a four-by-four matrix that summarizes the most important aspects of a program under consideration. The framework describes what is to be achieved and the means by which the achievements will be verified. The matrix is a summary presentation of the planned project or program. A logical framework is often part of a research project proposal, performance contract, or funding arrangement.

The logical framework was adopted by the US Agency for International Development (USAID) in the late 1960s. Since then, almost every international cooperation agency has incorporated the framework as part of its project cycle. The German technical cooperation organization, GTZ, for example, has combined the logical framework with new communication techniques and other analytical tools into a methodology called objective-oriented project planning (German acronym "ZOPP"). Today, several agencies are developing ways to combine the logical framework with participatory planning and evaluation procedures and refining the method further for project cycle management. The Consultative Group on International Agricultural Research (CGIAR) is using the logframe to consolidate and streamline its wide variety of activities towards common goals.

Logical framework approaches are most effective when they are applied in a collaborative mode. Stakeholder involvement enhances participants' commitment and project credibility, and teamwork taps the expertise of planning participants to the maximum extent. Participatory techniques like workshops and group facilitation are strongly recommended when using a logical framework approach.

Using logical frameworks

A completed logical framework matrix is the outcome of a planning process. It consists of guidelines to direct future action. Like every planning activity, it depends on solid analysis and negotiation through which project participants come to agree on the issues that the project will address. The quality of the project design largely depends on the inputs to the analytical process, specifically on the amount and quality of available information, the expertise and experience of the planning team, and the extent to which bias can be avoided and sources of creativity tapped. Careful choice of planning team participants ensures a good balance of expertise, qualifications, and interests.

Teamwork is a central element in most logical framework approaches. ZOPP, for example, is a participatory, workshop-based method. Planning teams are supported by a professional facilitator, who is responsible for helping the group meet its objectives. The facilitator guides the team by directing the flow of planning steps and regulating group processes. To exercise leadership, the facilitator should understand the problems at hand. But that person need not be a subject-matter specialist. It is best if he or she has no stake in the eventual implementation of the project.

Table 1 presents the essential structure of a logical framework matrix. While terminology may vary among organizations using the approach, the logic remains the same. That is, there are three causal links from a research program's activities to its ultimate goal: the link between activities and outputs, that between outputs and purpose, and that between purpose and ultimate goal. Thus, program *activities* are necessary for achieving *outputs*. *Outputs* are necessary for achieving the *purpose*. And the *purpose* contributes to achievement of the ultimate *goal*.

Table 1. Basic Structure of a Logical Framework

Narrative Summary	Verifiable Indicators	Means of Verification	Critical Assumptions
Goal			
Purpose			
Outputs			
Activities			

Goal, purpose, outputs, and activities should be carefully defined. By themselves, a research program's outputs are seldom sufficient to achieve the purpose and goal. Complementary programs or policies may be needed. The purpose and goal may also be affected by institutional factors and external conditions that are beyond the program's direct control. These factors and conditions are written in the *critical assumptions* column. The vertical logic in table 2, for example, asserts that if the six outputs are achieved and if integrated pest management (IPM) strategies for mandated crops are adopted by relevant countries for dissemination, then the short-term purpose will be achieved.

There are six basic steps involved in completing a logical framework matrix:

1. Begin with the narrative summary column and describe the goal, purpose, outputs, and activities involved in the program.
2. Define, for each level, the critical assumptions that must hold true in order for the objectives to be achieved.
3. Verify the program logic by checking the cause-and-effect relationships between activities, outputs, purpose, and goal while taking the critical assumptions into account.
4. Identify indicators by which research progress can be monitored and evaluated.
5. Define the means of verifying what occurs at each level, including what data is to be gathered, from what sources, and techniques to be used in data collection and analysis.
6. Review the logical framework periodically in the light of research progress and changing circumstances.

The narrative summary

The narrative summary is the first column of the logframe. It describes the program's logic, its underlying development hypothesis, and its different objectives, activities, and outputs. The *goal* is the ultimate objective to which the program contributes. Often it is outside the domain of research, for example, to increase poor farmers' incomes. The statement of *purpose* describes the desired outcome of the program, such as increasing farm production. *Outputs* are the program's more immediate aims – specific results for which the program manager is held accountable. An example is the release of a maize variety or training of a group of farmers. *Activities* are the actions undertaken to obtain outputs.

For long-lasting programs or projects, long-term and short-term purposes are often identified, as in the example in table 2. In this case, the short-term purpose is to have environmentally sound and economically feasible IPM packages available. These need not yet be adopted by users. The project outputs, therefore, are the development and testing of innovative techniques and the strengthening of the organizational foundation. For the subsequent phase of the

Table 2. Example of a Research-Oriented Project Logframe

Narrative Summary	Verifiable Indicators	Means of Verification	Critical Assumptions
Goal sustainable, environmentally sound horticultural production attained			
Purpose *long term:* identified packages adopted by users in the region *shorter term*: environmentally sound and economically feasible integrated pest management (IPM) packages available to users in the relevant countries	at least four IPM packages tested and delivered to users by April 1996	project documents	*assumption for the goal:* environmental conditions remain favorable for IPM strategies
Outputs 1. operational structure of the IPM regional project established 2. cooperation with relevant institutions for the implementation of IPM in pilot areas established 3. available IPM techniques tested in the region 4. IPM techniques developed and tested in the pilot areas 5. personnel of relevant institutions trained 6. role of gender adequately considered	1. increasing amount of IPM information processed by the Africa office by the end of 1996 2. at least two memoranda of understanding signed by end of 1994 and the rest signed by 1995 3. list of selected IPM techniques for the various crops available at the participating institutions; at least three techniques being tested by mid 1995 and at least two ready for dissemination by end 1996 4. additional bioecological data to solve at least two problems by end 1995 5. training programs carried out as planned and an increasing number of collaborators capable of conducting activities by end 1992 6. recommendations for considering gender-related aspects available for the second phase	1. project progress report for crop coordinators, financial report 2. memoranda of understanding 3. IPM performance data on tested techniques (on-station and on-farm) 4. project documents, reports, and institutional reports	*assumption for the purpose:* IPM strategies for mandated crops disseminated and adopted in relevant countries
Activities 1. 2. 3. …			*assumptions for outputs:* 1. countries nominate suitable crop coordinators 2. cooperating institutions show necessary commitment 3. …

Source: GTZ 1996.
Note: The project detailed here is entitled "Environmentally sound plant protection for vegetable and fruit production."

program, new outputs and activities are defined for achieving the long-term purpose. It is insufficient to express the purpose in terms of supply of knowledge or technologies. Rather, the program purpose should be set out in terms of beneficiaries' use of the knowledge or technology created.

At the level of outputs and activities, the amount of detail should not be exaggerated. Only essential activities and outputs should be stated so that program logic and the development hypothesis remain clear. For the same reason it is preferable for any one program to have only a single purpose.

Planners must also take care to distinguish between objectives and indicators. Objectives are typically rather general statements. Indicators add measurable detail to each. In table 2 for example, the fourth output states that innovative IPM techniques will be developed and tested in pilot areas. The corresponding indicator specifies what the planners felt would be sufficient evidence to prove that the output has been obtained.

The narrative summary is typically developed from the purpose down to the activities. It is important, however, for planners to ascertain that the program purpose and outputs will effectively contribute to the *goal*. Clear objectives are most likely to be formulated when they are phrased in the future perfect tense, as if they had already been achieved (e.g., "innovative IPM techniques developed and tested") rather than in the present or future tense, where they are stated simply as open-ended possibilities (e.g., "conduct research on IPM techniques" is less precise). Using future perfect tense, that is, looking from the goal backwards, makes it easier for planners to define what are relevant outputs and activities, link them in a causal chain, and assess the plausibility of the chain. In so doing, planners test the plausibility of the development hypothesis.

Assumptions

An assumption is a condition that must exist if the program is to succeed, yet which is not under its direct control. Assumptions, therefore, specify the circumstances under which the program logic and development hypothesis remain valid. In other words, they describe the conditions that are required if the outputs of the program are to lead to achievement of the stated purpose. Critical assumptions can be seen as the key threats to the program that exist in the external environment.

Assumptions should be stated as positive conditions, as in table 2. Frequently used assumptions in agricultural research programs are the existence of an effective extension system and the availability of external inputs in local markets. A good logical framework contains only a few critical assumptions. If there are too many assumptions, or if the assumptions are unlikely to bear true, a review of the program design may be needed.

High-risk assumptions are called "killer assumptions." An example of a killer assumption is that inputs such as fertilizer or credit for small farmers will be available, while the probability that the average farmer can access these inputs is very low. In such a case, the project is unlikely to contribute to the goal and should be redesigned. Eventually, the technology proposed may have to be altered or additional activities incorporated. Well designed projects include no killer assumptions.

Assumptions should be monitored during program implementation. Adaptation of the logical framework or action to influence external factors may be called for.

Indicators

Indicators specify how the successful achievement of the objectives will be measured or recognized. They are stated in terms of quantity, quality, and time (and sometimes also in terms of location). The fewer indicators the better. Indicators that reflect the central concerns of the program or project are more important than large amounts of detailed information.

Stakeholders and beneficiaries play key roles in defining indicators, which should then reflect the aspects of the project most important to them. Qualitative measures should be considered as well as quantitative ones. At the highest level, that of the goal, indicators define how new technology or knowledge will contribute to change the world. At the level of the purpose, indicators specify what changes are to be made in the actions of the target groups or, in other words, by whom and how widely the new technology or knowledge will be used.

Means of verification

The means of verification specify where and how information that validates the status of each indicator will be obtained. Subjective assessments from beneficiaries, stakeholders, peers, and other relevant groups are legitimate sources of such information. If no means of verification can be found, the indicator must be revised. Indicators and means of verification lay the basis for monitoring and evaluating program achievements. They must therefore be practical and cost effective, to ensure that monitoring and evaluating can actually be done.

Relevance for agricultural research

The logical framework can be used by agricultural research organizations to improve planning and management of research projects and programs. It can help planners focus research on clearly defined objectives and clarify the logic of cause-and-effect relationships between activities, outputs, purpose, and goal. The logical framework also helps to standardize terminology used in planning research. The completed logframe matrix is a consistent, highly transparent record that improves documentation, enhances communication among partners, and facilitates monitoring and evaluation. Preparing a logical framework forces planning groups to reduce the complexity of a program to its essential elements and make these elements communicable to those outside the planning participants. Finally, with the trend towards more project-related research funding, the logical framework becomes a useful tool for specifying research outputs and gaining research funds. In fact, many agencies that support research now require a logical framework to be part of all proposals submitted for funding.

However, the logical framework does have weaknesses. Like other planning methods, it results in only a limited interpretation of reality. The vertical logic assumes a fairly direct relationship between activities and outputs. Complex problems with nonlinear relationships and feedback loops are therefore ill captured by the analysis. To accommodate process-oriented and highly interactive undertakings, like agricultural research, logframes must be updated regularly. Moreover, the method handles quantitative indicators better than qualitative ones, and consensus is sometimes difficult to reach among participants on what constitutes acceptable performance and how it should be assessed.

Participatory development of logical frameworks is most useful for programs or projects that require input from several people, or groups of people, over a prolonged period of time (two to five years). In such cases, shared understanding of project purposes and outputs is essential. Moreover, the size of the project or program probably justifies the cost of participation. For smaller projects, the logical framework still contributes to improve internal logic. Even when it is undertaken by one person, he or she will gain from organizing the project ideas in the framework.

Integration with other tools

The logical framework provides guidelines for action. It summarizes, at a given point in time, the understanding and intent developed by a group of people, and it lays the foundation for future program management. It is an open design tool. That means that additional analytical and communication techniques are required for program development and management.

A planning team that aims to develop a sound research program should use the logical framework as a learning device rather than as a blueprint for program design. For problem definition and choice of potential solutions, it should be complemented by tools and methods such as surveys, analysis of statistics and secondary sources, ex ante assessment of benefits, participatory rural appraisals, and workshops.

Logical frameworks can be worked out for programs, projects, and components of projects. For management, it may be useful to divide a program's logical framework into logical frameworks for individual projects. The program can then be conceived as a portfolio of projects that share a common goal.

Example

The logical framework matrix shown in table 2 was developed for a regional project in East Africa. The underlying development hypothesis is that the adoption of environmentally sound and economically feasible IPM packages by users in the region will contribute to the goal of sustainable horticultural production.

Outputs and activities are shown only for the first phase of the project, three years. During this period, the short-term purpose should be achieved. The long-term purpose should be achieved by the end of the project. A new logical framework is to be developed for each new project phase.

The number of outputs is limited to six. This allows readers to quickly grasp the essence of the project, namely to build operational capacity, define collaborative relationships, develop and test available techniques, train personnel, and consider gender aspects explicitly. Few critical assumptions are defined.

Although no indicator for the goal and no assumption for its sustainability were included, the logical framework matrix is very clear about the achievements expected and how these are to be verified.

References and recommended reading

CEC. 1993. Project Cycle Management: Integrated Approach and Logical Framework. Brussels: Commission of the European Community, DG VIII, Evaluation Unit.
Discusses the use of the logical framework within the overall project management cycle. It is useful for understanding how certain planning features of the logframe, such as verifiable indicators, milestones, and means of verification are used later in project implementation, monitoring, and evaluation.

GTZ. 1988. ZOPP: An Introduction to the Method. Eschborn, Germany: Deutsche Gesellschaft für Technische Zusammenarbeit.
Integrates the logical framework into a wider set of tools for objective-oriented planning, including constraint trees and "GANTT" charts.

GTZ. 1996. Project Cycle Management (PCM) and Objective-Oriented Project Planning (ZOPP). Eschborn, Germany: Deutsche Gesellschaft für Technische Zusammenarbeit.
Similar to CEC 1993, this publication discusses the interaction between planning by means of a logframe and project management.

Farrington, J. and J. Nelson. 1997. Using Logframes to Monitor and Review Farmer Participatory Research. Agricultural Research & Extension Network Paper No. 73. London: Overseas Development Institute.
Shows how participatory research can be planned and reviewed by means of a logical framework.

McLean, D. 1988. The Logical Framework in Research Planning and Evaluation. Working Paper No. 12. The Hague: International Service for National Agricultural Research.

An early paper on the use of the logical framework as a tool for planning agricultural research. The paper provides clear and concise descriptions of the different items in the logframe and how they may be interpreted for agricultural research.

Schubert, B., U. Nagel, G. Denning, and P. Pingali. 1991. A Logical Framework for Planning Agricultural Research Programs. Manila: International Rice Research Institute.

Provides an example of how the logical framework has been used to plan a research program. It shows the logical framework's use at higher levels in the planning hierarchy.

Sartorius, R. 1996. The third generation logical framework approach: Dynamic management for agricultural research projects. *European Journal of Agricultural Education and Extension,* 2 (4): 49–61.

Discusses some recent experiences in the use of the logframe as a planning and management tool in agricultural research.

Chapter 26
Information Systems for Research Planning

Richard Vernon

To develop a plan, a view of the present is needed. The more informed that view, the better the plan is likely to be. The view of the present needed to plan agricultural research should include information on current resource allocations across commodity and noncommodity research programs and among the various regions of a country. Particularly in a large research organization, a well designed and well maintained information system is a powerful aid to providing such information at minimum cost to those making planning decisions. But attaining a well designed and maintained information system is no trivial task. Trial and error has yielded lessons to guide future information systems development, to enhance the chance of their success. This chapter looks at information systems and how they are useful in research planning. It then looks at how such a system is developed, implemented, and operated.

What is a management information system?

A management information system (MIS) is a set of formalized procedures to provide managers, researchers, and sometimes other stakeholders with information based on data from internal and, if desired, external sources. Such information enables these users to make timely and effective decisions for planning, directing, and controlling the activities for which they are responsible.

While it might be possible to find the information needed to make a particular decision without recourse to a formal information system, it is likely to take considerable time and effort. An advantage of maintaining an information *system* is that it can hold all the information likely to be needed for routine decisions so that it is instantly available when called for: a decision does not have to wait for relevant data to be collected. Another advantage is that information collected and stored only once can be used in a number of management processes.

Typically, an MIS uses computer software known as a database. While an MIS does not have to be computer-based, without the use of modern information technology only a limited amount of data can be incorporated. Whereas word processing software manages primarily text and a spreadsheet is used mainly to store and calculate with numeric data, database software stores facts and figures and relations between the two. An MIS often holds very large numbers of records, for example, of experiments (their titles, objectives, and results) and scientists (names, education, and discipline). The MIS extends the functionality of a database by providing a range of built-in standard reports, tailored printable forms for use in data capture, and matching on-screen forms for easy data entry. Its menu system enables users with very little knowledge of the underlying database or the theory of databases, to select and access information and reports as they wish.

Information needs for agricultural research planning

Managers of agricultural research require information relevant to each stage of the research cycle: planning, implementation (monitoring research projects), and delivery of outputs such as new varieties and recommendations for farmers and scientific publications. Information needed for planning therefore relates to four main components:

- inputs (resources) available, including personnel, finance (budgets), and physical resources (land, laboratories, equipment)
- the current research program and its projects
- policy issues, such as priorities across commodity topics (e.g., maize, bananas, goats) and noncommodity subjects (e.g., soils and economics)
- external factors that have a bearing on the research program, for example, commodity prices and import and export data

An MIS not only stores and presents such information in isolation; it also links information, which facilitates the planning processes. For example, it can link information on scientists and the commodity focus of research projects with declared priorities (see figure 1).

Human resource information needed for planning includes a basic inventory of all research staff together with some basic details, as shown in table 1.

Financial information useful for planning may include a broad statement of expenditure for the previous year as a basis from which to develop a budget for

Table 1. Information on Scientists Contained in an MIS

Personal	Administrative	Scientific	Current Program
Name	ID number	highest degree	list of projects and time spent on each
Date of birth	rank	scientific discipline	
male/female	job title	main commodity (crop or animal) or noncommodity focus	
		location	

the new year. A powerful MIS feature is that if expenditure has been recorded at the project level, the information system can aggregate it easily to the level of parent programs or research stations and also show allocations across commodities, scientific disciplines, and agroecological zones.

Planners also need information on the research program itself. Details are recorded on all projects and experiments and how they are grouped into programs and departments. Information on the structure of research facilities, such as locations and capacities of research stations and institutes, is also useful. Finally, planning has to take account of external factors such as national objectives that may be found in five-year national plans, levels of food imports and exports, and the current status of the labor market.

These several classes of information needed for research planning can be housed within an MIS. Some categories, such as finance, personnel, and physical resources, are important for administrative reasons other than research program planning. For these functions, separate systems (accounting, personnel) are usual, but these may be linked to the research MIS to enable single data capture and storage. It should also be recognized that some information does not lend itself to capture in such a structured system. Yet, such informal information may also influence management decisions.

Developing and using an MIS for planning

Information systems are usually costly to develop and implement. Moreover, experience shows that they often fail to deliver benefits commensurate to their costs. Yet that same experience has also yielded a list of some common causes of failure. From this list, a set of critical factors and lessons for success can be derived. These relate to the system itself, its implementation, and its institutionalization.

The information system

To be useful for planning, an information system must conform to an agreed level of quality, particularly in its content and ease of use. Regarding the information content of the system, three attributes should be considered:

- *Relevance.* It is common to find information that is collected but not used. This wastes collection effort, but more seriously, distracts users (e.g., managers) from more important facts and figures.
- *Accuracy.* A minimal level of accuracy is needed, but too high a level is wasteful. For example, figures on resource allocations across commodities usually need not be more accurate than within, say, 10 percent of actual.
- *Currency.* The value of a lot of information declines with time. An information system needs swift data capture, processing, and output subsystems for its reports to be of maximum value. In an agricultural MIS a maximum turnaround time of a few weeks is a realistic target. On the other hand, periodicity

or frequency of data capture can be exaggerated: most experiments have an annual cycle. It should thus be sufficient to capture most data related to an experiment once a year.

Implementation

The system has to be closely integrated into the research organization's planning and management cycle (see Horton and Dupleich, this volume). Timing is important, so that information needed for a particular planning meeting is captured and processed in time for reports to be available at or before the meeting.

The system should produce useful outputs soon – within days or, at the most, a few weeks after data capture. Further, such outputs must be made available to all the parties involved. Prompt and wide distribution of reports provides powerful motivation for those supplying the data. Early availability of information outputs is also essential to facilitate the next round of project or experiment planning.

Institutionalization

To integrate the MIS fully into the organization's planning and management cycle, the MIS needs a "patron" who is a member of top management and who is convinced of the value of such a system. The patron ensures that outputs of the MIS are demanded and inputs provided. Those who use the system, such as research station managers and program leaders, and those likely to be most affected by it, such as scientists, must be consulted during MIS development and implementation. Nowadays, many information practitioners advocate implementation of a "prototype": a simple system with just the main features that users have asked for. A prototype can be developed in a short time, perhaps just a few weeks, and passed to users for testing.

Relevance for agricultural research

The essential advantage of a management information system is that with a basic set of data on research activities and scientists, collected or updated annually, a wide range of outputs can be produced to inform each stage of the planning process. The MIS can provide a snapshot of the current distribution of funds and scientists across, for example, the organization's priorities (figure 1), commodities (figure 2), agroecological zones, and scientific disciplines.

Comparison of current resource allocation across commodities with agreed upon priorities provides valuable inputs to early planning discussions. There are nearly always significant mismatches. Three issues then arise. The first is the validity today of the ranked list of priorities. Priorities are usually set at intervals of several years. Any subsequent debate has to assume that priorities are still

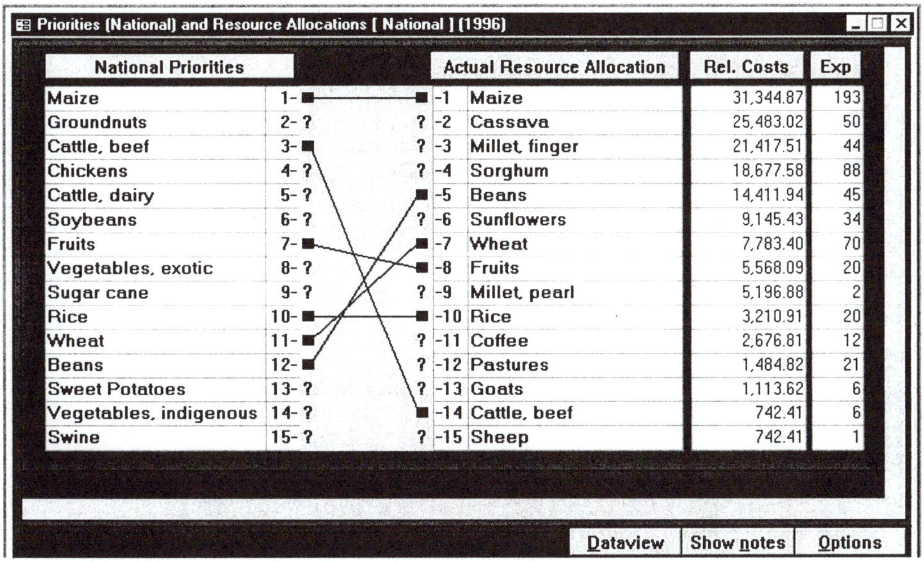

Priorities (National) and Resource Allocations [National] (1996)

National Priorities				Actual Resource Allocation	Rel. Costs	Exp
Maize	1- ■	■	-1	Maize	31,344.87	193
Groundnuts	2- ?	?	-2	Cassava	25,483.02	50
Cattle, beef	3- ■	?	-3	Millet, finger	21,417.51	44
Chickens	4- ?	?	-4	Sorghum	18,677.58	88
Cattle, dairy	5- ?	■	-5	Beans	14,411.94	45
Soybeans	6- ?	?	-6	Sunflowers	9,145.43	34
Fruits	7- ■	■	-7	Wheat	7,783.40	70
Vegetables, exotic	8- ?	■	-8	Fruits	5,568.09	20
Sugar cane	9- ?	?	-9	Millet, pearl	5,196.88	2
Rice	10- ■	■	-10	Rice	3,210.91	20
Wheat	11- ■	?	-11	Coffee	2,676.81	12
Beans	12- ■	?	-12	Pastures	1,484.82	21
Sweet Potatoes	13- ?	?	-13	Goats	1,113.62	6
Vegetables, indigenous	14- ?	■	-14	Cattle, beef	742.41	6
Swine	15- ?	?	-15	Sheep	742.41	1

Dataview | Show notes | Options

Figure 1. Comparison of agreed priorities with actual resource allocations across commodities

valid. If the planning team feels that adjustments are due, these are best agreed upon before further comparisons are made with resource allocations. Once the team is comfortable with the priorities, it must plan how resource allocation can best be adjusted to bring it more in line with priorities. This is usually a gradual process, done at annual research planning and budgeting meetings at which figures and reports come into play (see Bruneau, this volume).

Finally, some attention should be given to how mismatches came about and what can be done to minimize them in future. Usually they are the product of years of program development with insufficient regard for priorities set. Some years may then be needed to bring the two back in line. Sometimes funding agency priorities cause imbalances between organizational priorities and resource allocations. In these cases, an MIS provides powerful food for thought to financiers and stakeholders.

Comparison of resource allocations to the major commodities with priorities is typically the task of a planning team. At the organization and station levels, a more detailed view of commodity support is provided by a report of scientists' time allocated to the main commodity groups, such as oilseeds, cereals, and livestock. These proportions can then be compared with the organization's or station's mandate. Any significant differences should be addressed in the research planning process, at which time resources can be redistributed to "underweight" commodity groups and factors. In a second stage of this process, scientists' time is examined in relation to individual commodities. An MIS re-

Researchers Time Allocation to Commodities [Maastricht] (1996)

Commodity	Person Years	Researcher Cost
Maize	3.8	31,344.87
Cassava	1.9	25,483.02
Unknown	2.2	23,455.38
Millet, finger	2.3	21,417.51
Sorghum	2.3	18,677.58
Beans	1.2	14,411.94
Soils	1.5	11,047.64
Sunflowers	1.0	9,145.43
Wheat	1.0	7,783.40
Many	0.7	7,427.59
Citrus	0.8	5,568.09
Millet, Pearl	0.7	5,196.88
None	0.4	4,665.32
Rice	0.3	3,210.91
Coffee	0.3	2,676.81
Totals:	21.7	202,002.55

Ascending

Record: 1 of 35

Maastricht | Dataview | Show notes | Options

Figure 2. Human resource allocations to commodities

port for this is shown in figure 2. Other management issues may arise at this level. For example, a commodity with very low total researcher input (perhaps less than 10 percent of a person year) might raise questions about the likelihood of this work yielding useful outputs. It may be permitted for initial exploratory work in a new area. But if such small input has persisted for several years the limited resources may be better invested in maintaining a literature survey of research being done elsewhere on the topic .

While other such management reports can be obtained from the MIS, the more basic reports often prove most valuable, such as an inventory or listing of all experiments by commodity and by station. Using keywords as descriptors, specific research types, themes, or thrusts can be explored, for example, current and planned work on integrated pest management or post-harvest technology.

A similar set of outputs can be produced from data on scientists. Any research program is dependent on its cadre of researchers. A primary MIS output is thus a directory of scientists with details of their disciplinary and commodity focus. An information system can easily provide this sorted by commodity, discipline, and station in addition to the usual "by last name" or "personal identification number" sort orders. This is yet another example of the same underlying data presented by the MIS in different forms to suit different management needs.

Numbers of scientists working on each commodity, in each agroecological zone, and in each discipline are also useful reports. Various age-related issues can be explored by looking at numbers of staff likely to be retiring soon from each commodity or discipline: the system can generate this information from the base data on scientists' dates of birth (see table 2). This information might advise training and recruitment programs.

When a research project is first proposed, its details can be entered into the MIS and, in aggregation with other proposals and the existing ongoing program, implications on available resources and resource distribution explored. When a decision is made on each proposal, accepted projects are recorded as approved, which is a point of closure. No further discussion is entertained and the project can proceed to implementation.

Upon completion of a project, the MIS provides a straightforward tool for its evaluation, matching the project's outputs with its original objectives as recorded in the system. Disparities provide material to inform the next round of planning.

An MIS has several costs of which the software is only one: hardware and the training that is part of institutionalizing the system can be expensive. Several questions can be asked when considering implementing such a system: Will a ready-made MIS be adopted and adapted, or should a new system be designed from scratch? The latter can be expensive, take years, and offers no guarantee of success. What resources are needed to support the project? Will the system be developed within the organization or will the task be outsourced to a commercial enterprise? Some of the references given below will assist in finding the answers to these questions.

Table 2. Current and Projected Numbers of PhD Holders by Discipline

Discipline	Number of PhDs	Retiring by Year 2005	In Training	Potential Number in Year 2005	Percent Change
Plant production	12	3	4	13	8
Plant protection	8	1	3	10	25
Soils	7	2	4	9	29
Animal health	5	1	0	4	–20
Plant breeding	5	2	2	5	0
Socioeconomics	3	0	3	6	100
Postharvest	2	0	0	2	0
Animal production	2	1	3	4	100
All	44	10	19	53	20

Note: The same can be done for MSc and BSc graduates.

Recommended reading

Laudon, K. C. and J. P. Laudon. 1994. Management Information Systems. New York: Macmillan Publishing Company.

McNurlin, B. C. and R. H. Sprague (eds). 1989. Information Systems in Practice. Second Edition. Prentice-Hall International.

Vernon, R. (in press). Information Systems for Agricultural Research Management. The Hague: International Service for National Agricultural Research.

ISNAR's Internet site, http://www.cgiar.org/isnar/ offers a demonstration of the management information system "INFORM-R" and a facility for downloading the INFORM-R software.

Chapter 27
Participatory Rural Appraisal

Vanessa Henman and Robert Chambers

There has been an explosion recently of methods to enable farmers to express, present, and analyze their knowledge and to share this with scientists and extensionists. Many of the methods evolved from agroecosystem analysis and entail farmers making observations, maps, and diagrams. These are now described as "participatory rural appraisal" (PRA) methods. PRA is an extended process of appraisal and analysis that can lead to local action by a community or group. Crucial to the successful use of these methods are the attitudes and behavior of the facilitator. PRA is particularly useful in fostering interaction with resource-poor farmers.

What is participatory rural appraisal?

Participatory rural appraisal (PRA) is a method used by researchers who want to plan their work in close collaboration with a rural community. PRA is a process of appraisal and analysis that may lead to local action by a community or group, such as the establishment of farmer-planned and designed experiments. PRA not only improves researchers' information on farming constraints; it also improves interaction and exchange of ideas between rural communities and researchers. Moreover, the logical conclusion of a PRA process is the joint planning of research projects or experiments (see also Sperling and Ashby, this volume). The PRA tool kit is diverse and growing. It combines instruments for information sharing with tools for collective decision making and planning.

PRA does not necessarily or exclusively lead to joint planning and execution of research projects. While undertaking a PRA, problems may emerge that need to be investigated at experiment stations or in the laboratory. But for those problems that can be handled within the rural locality, research plans are drawn up together with farmers. Participation of farmers in planning and executing research tends to increase the relevance of results and the likelihood of achieving outcomes that can be applied in the rural context. PRA is especially useful when research is aimed to help resource-poor farmers who work under severe financial and ecological constraints.

An approach closely related to PRA is rapid rural appraisal (RRA). Both offer creative means of information sharing and challenge prevailing biases and preconceptions about rural peoples' knowledge. Whereas RRA is a more extractive, eliciting approach in which the main objective is data collection by outsiders, PRA recognizes that besides producing timely and relevant knowledge, rural people should have control over the use of information that they provided or helped to collect (Waithaka 1998).

Doing participatory rural appraisal

There are no strict rules for doing PRA. There are four points, however, that should be kept in mind: choice and sequencing of methods, selection of farmers, triangulation of data, and behavior and attitudes of facilitators.

Choice and sequencing of methods

An extensive range of PRA methods is available, including older techniques like joint observation, farm walks, transect walks, semistructured interviews, and discussion in focus groups of farmers. Newer, visual methods have proved powerful for the presentation, analysis and discussion of complex farming realities. These can complement some of the older methods. Visualization often entails drawing and diagramming on the ground or on paper with sticks, chalk, powders, or pens. Or it may involve sorting items, cards, or symbols and scoring and estimating using local materials such as stones, seeds, and lengths of sticks or straws. These enable detailed and sophisticated analysis, often at a level impossible to achieve through discussion alone. Visualization makes it easy for farmers to add to and modify information, progressively elaborating on the knowledge shared. Visualization, furthermore, encourages wide participation by enabling less confident and illiterate community members to express their views visually. To express farmers' reality in all of its complexity, PRA methods may combine various dimensions (table 1).

Of the rich diversity of methods available, some have proved especially fitting for encouraging interaction between farmers and agricultural scientists (see table 2). PRA methods can be used in conjunction with traditional research

Table 1. Dimension and Methods of PRA

Dimension	Methods
Spatial	mapping and modeling
Nominal	collecting, naming, listing
Temporal	sequencing over days, seasons, or years
Ordinal	sorting into types, ranking
Numerical	counting, estimating, scoring
Relational	linking to show flows and connections

Table 2. Tested Methods for Use by Farmers and Scientists and Examples of Applications

PRA Method	Examples of Applications
Seasonal calendars	examine seasonal patterns in the incidence of animal or crop pests and diseases, rainfall, household expenditure, or farm labor
Venn diagramming	reveal the importance, relevance, and involvement of local and external institutions in addressing agricultural issues; examine sources of new agricultural ideas and information and determine partners for work
Timelines and trends	enable analysis of change over time, such as in crop varieties grown, extent of soil erosion, occurrence of drought, herd numbers, diet
Matrix scoring	examine peoples' own criteria for choosing among options such as crop varieties, soil fertility measures taken, and characteristics of a good irrigation system; indicate the severity of problems such as animal disease
Resource and agroecological zone mapping	map areas sharing similar characteristics in terms of soil types, crops grown and rotations, and land access and tenure
Causal and impact diagramming	show flows, causal relationships or other connections, such as expected impact of an irrigation system or the causes of soil erosion
Farm mapping and flow diagramming	map individual farm plots and their location in relation to each other; examine different soil management practices according to distance from the homestead, crops grown, and rotations; examine nutrient flows within the farm system or the division of labor at the households level
Transect walks	learn about the locality, crops and trees grown, soil types, and amount of fallow land
Farm type sorting, wealth and well-being ranking	learn the ways in which people differ with regard to wealth or well-being and gain a quick understanding of relative socioeconomic status and households' and the community's definitions of wealth; assist in identifying key informants for other PRA exercises

methods: the way in which each complements the other is increasingly recognized (Abbot 1997). For example, farmers and scientists can use agroecosystem and farm mapping to identify locations where soil samples should be taken (Turton et al. 1997) or to find particular types of farmers to be interviewed.

Flexibility is key in selection and use of PRA techniques. Methods should be developed and evolve to meet particular circumstances. Mistakes in method selection should not be seen as failure. Rather, they are trial-and-error learning opportunities from which subsequent exercises can be refined. As such, new forms and combinations of methods are continually being devised. Among the many ways in which information is expressed are maps, models, matrices, pie diagrams, card piles, lists; ranks, scores, histograms, and graphs, as well as spider, causal, linkage, and Venn diagrams.

There seems no limit to visual and diagramming inventiveness. For example, mapping has now been modified and adapted beyond village-scale mapping to include, cascade mapping of irrigation systems spanning many villages and catchment-level mapping for understanding soil and water conservation issues. At the other end of the scale, body mapping of animals has been invented to investigate farmers' perceptions of pests and diseases affecting livestock and their remedies. Other innovations include mobility mapping, to investigate where people go for particular resources or to obtain new agricultural ideas.

As with the choice of methods, no blueprint exists for the order in which PRA exercises should be done. PRA frequently begins with mapping or timelines, as these impart general information about an area. Then it might be appropriate to move on to methods such as Venn diagramming or matrix scoring. However, local circumstances should be considered in choosing methods. For example, mapping may not be appropriate in sensitive areas such as those adjacent to protected areas or where there are or have been boundary disputes. While it is useful to start with a plan, it is equally important to be flexible to enable methods to flow from one to another according to information arising at each stage. This approach has the further advantage of enabling practitioners to avoid becoming too mechanical in the process (for an example of sequencing, see Turton et al. 1997).

Which farmers

Farmers' perceptions, daily realities, needs, priorities, and opportunities will vary according to many factors. Some of the most significant are one or a combination of gender, wealth, age, and ethnicity. Hence, it is always important to know which farmers are involved in PRA exercises. The tendency for them to be men rather than women, and better-off rather than poorer, needs to be resolutely countered in order to gain a balanced view. However, it is also important to avoid making assumptions about different groups and creating artificial groupings according to presumed differences. The differences that affect people's livelihoods in the communities in question are the ones that should be examined (Cornwall 1998). Methods such as farm sortings and wealth and well-being rankings can be used to help identify these categories.

Some exercises are best conducted in groups and others with individuals. Groups tend to be good in exploring general issues and are often most useful at the start of the research process, although their input may also be valuable at decision-making stages when it is important for all stakeholders to be involved. In some instances, PRA exercises might need to be done with different groups, such as men's and women's groups, simultaneously. Each group can then present its results to the others, stimulating debate and enhancing understanding. PRA conducted at the individual or household level may be appropriate for exploring specific issues, or those of a sensitive nature.

Triangulation of data

While PRA stresses the magnitude and complexity of local peoples' knowledge such knowledge should not be accepted at face value. Factors such as farmers' past experiences with researchers, the prevailing development rhetoric, government policy, and expectations of the organizations involved all influence local people' perceptions (see Christoplos 1995, Lindblade 1997). Moreover, researchers themselves have biases and the potential exists for misinterpreting PRA results. For example, in Kabale, Uganda, during historical mapping exercises, local people reported that peas were no longer grown owing to soil infertility; yet peas were observed growing on a considerable number of plots during transects. The reality was that peas were no longer cropped alone as they had been in the past (Lindblade 1997). Rather than being a weakness of the approach, discrepancy and contradictions enable the PRA team to learn more about the issue under study. Discrepancies should be seen as opportunities to learn further rather than as errors to be glossed over or ignored. In such cases, the PRA team should return to the sources of any apparently contradictory information and probe further (Nabasa 1995).

Triangulation or cross-checking of data is an important but often forgotten component of participatory work and can take various forms. The use of multidisciplinary teams, using the same methods but with different informants, and investigating the same theme using different research methods can all help determine the validity of data collected.

Behavior and attitudes

In PRA, the behavior and attitudes of outsiders – scientists or extensionists in the case of agriculture – have proved more important than the methods used. Outsiders come as facilitators, not teachers. When outsiders dominate, farmers are inhibited or reflect back what was said. To empower farmers to freely reflect, present, and analyze, the outsider has to learn to be low key – to initiate a process and then sit down and keep quiet for much of the time, to listen and watch. These behaviors do not come easily to scientists and extensionists who are energetic, enthusiastic, and knowledgeable. Their normal behavior has to be unlearned and new, less forceful, lower profile behavior and attitudes adopted. Only then can PRA methods be well facilitated and their potential realized. A number of tips can assist in PRA's application:

- Have a team contract. Agree among yourselves who will be the main facilitator and that others will not interrupt.

- Take time to develop rapport. Explain who you are, why you have come, and what farmers can and cannot expect from your visit. If they can expect nothing, make that very clear. Unless they are very busy, farmers are usually willing to take part and will find the activities interesting.
- Relax, do not rush. Allow more time than you expect you will need.
- Hand over the stick. That means pass the initiative to farmers. Do not do anything for them (e.g., counting out seeds or drawing a matrix) that they can do for themselves. Show confidence that farmers can use PRA methods.
- Remember that the key output is not the matrix, map, or other end product, but rather the discussion and analysis among farmers that led to the product's development.
- Keep quiet. Do not interrupt, criticize, or put forward your own ideas or knowledge until the end. This is extremely difficult. But the more researchers share their knowledge, the more farmers will defer to them and more difficult it is for them to express their own thoughts.
- Interview the map or diagram once it has been completed. There is often much to be gained from discussing it. It provides an agenda. Farmers are usually very willing to explain what they have shown and why.
- Do not "convert" local classifications and terminology into "scientific" terms, as this may undermine local knowledge. Try instead to learn as much as possible from the criteria used for local classifications.
- Remember that the information generated belongs to the farmers involved. Discuss how it is to be used and what records should be kept where. If results are to be published, request permission from the farmers and give due acknowledgment.

Relevance for agricultural research

PRA arose from concerns about the quality of data collected using traditional questionnaire surveys. Much relevant information falls outside the scope of such questionnaires. That which is not accommodated is usually ignored. One strength of PRA is that it enables researchers to take local priorities into account in their research by enabling expression of local complexity. Results gleaned using conventional research methods are often quite different from those generated using PRA. For this reason, farmer participatory approaches are increasingly being adopted to pinpoint researchable issues and develop on-farm research plans for the selection and breeding of crop varieties and for other types of agricultural research (Witcombe 1996).

PRA methods have frequently revealed considerable differences between farmers and researchers in the criteria they use to select and adopt new technologies. For example, in India when a matrix scoring of wheat varieties was done separately by a researcher and farmers, the researcher selected seed primarily based on characteristics of high yield and resistance to insects and pests. Farmers placed importance on a whole other range of factors. Seed that did not shatter was more important to farmers than resistance to insects and pests, yet this characteristic did not even feature on the researcher's list of criteria. Implications for research planning obviously are great.

PRA techniques are most commonly used in the initial stages of the agricultural research process in problem identification and analysis and in exploration of possible solutions. Yet PRA methods can be used at all stages in agricultural research, by farmers and researchers, including in the planning and design stage of experiments and in the monitoring and evaluation of both on-farm and on-station trials. Through participatory processes, scientists boost the own-initiated research in which farmers are already involved as well as supporting the work of local institutions. PRA thus facilitates the development of research plans that are linked to the overall development requirements of rural localities.

Example

In a study covering 13 villages in Nepal, a variety of PRA and more conventional research methods were used by a multidisciplinary team of researchers from a government research organization. The aim was to explore the complex issue of soil fertility in the hills of Nepal. The sequence of methods used for initial problem definition and analysis is shown in figure 1. Use of conventional surveys in parallel with participatory approaches was found to be one way to

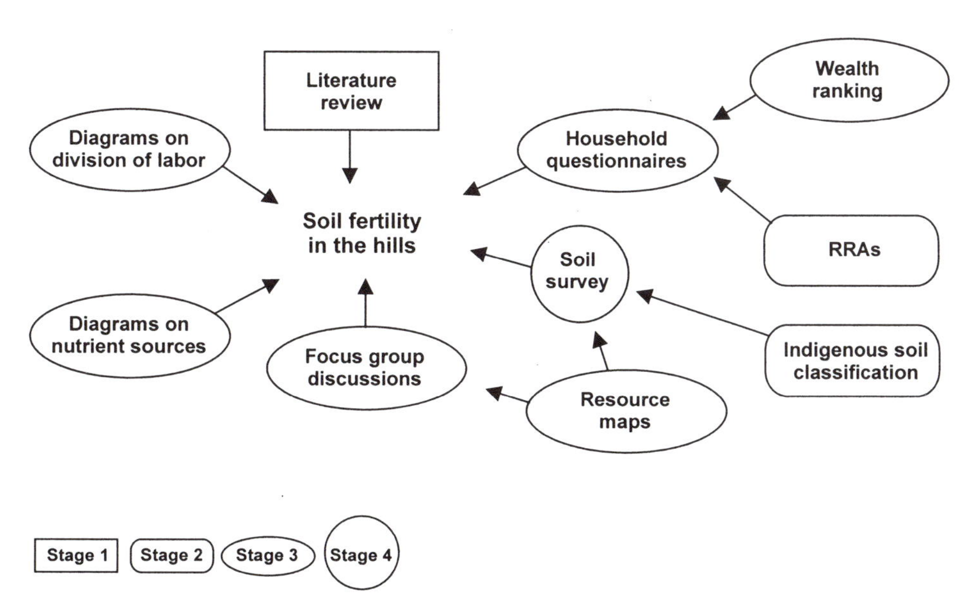

Figure 1. Sequence of methods used in a participatory rural appraisal

achieve "breadth of coverage" whilst maintaining the "depth and quality" of the information obtained (Turton et al. 1997).

References

Abbot, J. et al. (eds). 1997. Methodological Complementarity. PLA Notes No. 28. London: International Institute for Environment and Development.

Bainbridge, V. 1998. Participatory Approaches and Agriculture Topic Pack. Brighton: Institute of Development Studies.

Chambers, R., A. Pacey, and L. Thrupp (eds). 1993. Farmer First: Farmer Innovation and Agricultural Research. London: Intermediate Technology Publications.

Christoplos, I. 1995. Representation, Poverty and PRA in the Mekong Delta. Research Report No. 6. Linköping, Sweden: Linköping University.

Conway, G. 1985. Agroecosystem analysis. *Agricultural Administration,* 20: 31–55.

Cornwall, A. 1998. Gender, participation and the politics of difference. In *The Myth of Community: Gender Issues in Participatory Development* edited by I. Guijt and M. Kaul Shah. London: Intermediate Technology Publications.

Drinkwater, M. 1993. Sorting fact from opinion: The use of a direct matrix to evaluate finger millet varieties. *RRA Notes,* 17: 24–28.

Guijt, I. and J. Pretty (eds.). 1992. Participatory Rural Appraisal for Farmer Participatory Research in Punjab, Pakistan. London: Sustainable Agriculture Programme.

Jones, C. 1997. PRA Behavior and Attitudes Topic Pack. Brighton: Institute of Development Studies.

Kumar, S. (ed.). 1996. ABC of PRA: Attitude Behavior Change. Patna, India: ActionAid. (Available on request from the Institute for Participatory Practices PRAXIS, 12 Patliputra Colony, Patna, Bihar 800 013, India. e-mail: praxis@ActionAidIndia.org.

Lindblade, K. and G. Carswell. 1997. Discrepancies in understanding historical land use changes in Uganda. In *Developing technology with farmers: a trainers guide for participatory learning* by Van Veldhuizen et al., pp.59-63. PLA Notes 28.

Manoharan, M. K., K. Velayudham, and N. Shunmugavalli. 1993. PRA: An Approach to Find Felt Needs of Crop Varieties. *RRA Notes,* 18: 66–68.

Nabasa, J. et al. 1995. Participatory Rural Appraisal: Practical Experiences. London:: National Resources Institute.

Paliniswamy, A., S. R. Subramanian, J. N. Pretty, and K. C. John (eds.). 1992. Participatory Rural Appraisal for Agricultural Research at Paiyur, Tamil Nadu, Centre for Agricultural and Rural Development Studies. Coimbatore: Tamil Nadu Agricultural University and London: International Institute for Environment and Development.

Premkumar, P. D. 1994. Farmers are Engineers: Indigenous Soil and Water Conservation Practices in a Participatory Watershed Development Programme. Pidow-MYRADA, Prakruthi-Gnana Kendra, Kamalapura 585313, Gulbarga District, Karnataka, India.

Turton, C. et al. 1997 . The use of complementary methods to understand the dimensions of soil fertility in the hills of Nepal. *PLA Notes* 28 (Feb): 37–41.

Waithaka, M. 1998. Integration of a User Perspective in Research Priority Setting: The Case of Dairy Technology Adoption in Meru, Kenya. Kommunikation und Beratung No. 22. Weikersheim: Margraf Verlag.

Witcombe, J. 1996. Participatory approaches to plant breeding and selection. *Biotechnology and Development Monitor,* 29 (Dec.): 2–6.

Further sources of information

Institute of Development Studies (IDS), University of Sussex, Brighton BN1 9RE, UK Tel: +44 1273 877263 Fax: +44 1273 621202.

International Institute for Environment and Development (IIED), 3 Endsleigh Street, London. WC1H ODD. Tel: +44 171 388 2117.

Chapter 28
Alternative Scenarios for Agricultural Research

Bruce Johnson and Maria Lucia D'Apice Paez

Alternative scenarios describe future conditions in which an organization (or system of organizations) may have to operate, as defined by sets of distinct hypotheses on key variables that affect the development of the organization. Only a limited number of scenarios is usually developed. Scenarios are built around variables that are independent of the organization but that strongly affect its functioning and position. The hypotheses underlying each scenario need to be plausible, consistent, and relevant. In situations of complexity and rapid change, scenarios represent explicitly the interrelated uncertainties that are most important to a planning problem at hand. The use of scenarios, as an adjunct to the planning process, provides useful insights about an uncertain future and improves perceptions and judgments in decision making.

What are alternative scenarios?

Scenario development is a way of generating relevant information about an uncertain future. Agricultural research organizations may apply scenario development in their long-term planning processes, especially when they observe that the socioeconomic and political environment in which they operate is starting to change (see also Rutten, this volume). Scenario development is a disciplined method for imagining, structuring, and analyzing possible futures. It is being applied in a growing number of government and private organizations. The result of scenario development is usually a small set of alternative scenarios that highlight and contrast the different conditions that research organizations may expect to face. Confrontation with these contrasting scenarios tends to improve decision makers' perceptions and judgments in strategic planning.

Agricultural research must necessarily work with long time horizons, given the natural rhythm of biological processes. Plant-breeding programs, for example, may take 10 to 15 years of focused effort to develop new varieties. Yet institutions, funding, research methodology, and agricultural markets are subject to change over such long periods. Planning must therefore take into consideration not only present but also future conditions so as not to jeopardize institutional sustainability. If demands for new technology are not met in a timely fashion, if scarce resources are dedicated to problems of declining priority as a result of market changes, or if vital research programs are discontinued or hampered due to administrative misunderstandings, planning has failed due to its incapacity to anticipate changes in the external context.

Initially developed by the "Rand Corporation" for geopolitical studies, scenarios are now widely used to counteract the "tunnel vision" of understanding the future strictly in terms of the past and present. They are used in formulating long-term policy, institutional strategy, and research programs. Alternative scenarios can play an important role in planning by bringing into sharp focus a wide view on the effects of changes in the socioeconomic and political landscape. Scenarios developed with the participation of researchers and beneficiaries bring the added benefit of promoting the internal changes that may be required, as the external need for change is made explicit.

Scenarios are developed by focusing on a limited number of *critical issues* in the planning problem. These involve forecastable and unforecastable changes in the context that appear to have the greatest potential impact on the functioning and position of the organization. Scenarios are developed to reflect a *probable* (or expected) future as well as *possible* (or alternative) futures. For example, change in agriculture policy from an export-oriented, centrally planned system to a highly diversified and decentralized market-driven sector may be considered improbable over the planning horizon. The impact of such a change on agricultural research priorities, however, would be dramatic and needs to be contemplated in the planning process to ensure institutional and research program sustainability.

Each critical issue is examined to identify the underlying causes or driving forces that can influence future outcomes, and the interrelations among critical issues are mapped. Trends that are identified in this manner are combined with hypotheses concerning future events (e.g., a change in the type of government or in the import policies of a major trading partner). Alternative scenarios can then be seen as different combinations of possible trends and hypotheses. A common set of variables, while potentially capable of generating an almost infinite number of combinations, is used to create a limited number (usually two to five) of alternative scenarios. These alternative scenarios have different implications for the role and position of the agricultural research organization or system. Each tells a future in which the same elements interact under different conditions, producing distinctive results.

Scenarios are based on systems analysis of underlying causes and effects, to produce a holistic vision of the future. To obtain the desired utility for decision makers who, after all, must place themselves in the diverse situations described and perceive implications in terms of their future actions, scenarios must be plausible, internally consistent, and provide information that is relevant to the planning and decision questions at hand.

Developing and using alternative scenarios

A number of methodologies for scenario planning are described in the literature. All of them present a sequence of steps to be followed in elaborating and using scenarios, with variations due to special purposes or circumstances. A few activities are common to most of these methodologies, and these are detailed here.

Definition of scope

The objectives of scenario development must be clearly defined in terms of time horizon, purpose, and width and depth of scope. These elements are strictly tied to the planning and decision question being deliberated. If the problem to be addressed is the formulation of a national research policy, the scenarios should encompass a long-term view of at least a decade, with a correspondingly wide scope looking at the agricultural sector and its relationship with society as a whole. If the problem is development of a research organization's institutional strategy or master plan, the time frame should be shorter, generally five to 10 years. The scope would also be narrower, but with greater depth in the areas of most concern to the organization, such as governance, funding, research strategy, priorities, and linkages. Scenarios may also be used in planning some research programs and projects, for example, those focusing on agroindustrial production chains, regions, or ecosystems.

Identification and analysis of critical issues

Within the scope already defined, critical issues are identified and prioritized with the involvement of both internal and external stakeholders. Initially an extensive list of critical issues is generated. This list is then subjected to critical review and consolidation. Finally, a selection is made of a limited number (generally 10 to 15) critical issues of greatest potential interest to stakeholders and of greatest impact on the planning question. The scenario team then identifies the basic political, economic, social, technological, legal, and institutional trends that underlie the selected critical issues. Each trend is analyzed, in external consultations if necessary, in order to grasp the causal factors driving or restraining the trends. Depending upon the circumstances, open hearings or structured debates may be organized to enrich understanding of the complexities involved. The purpose of the analysis is to understand the causes of uncertainty and anticipate possible future outcomes for each critical issue.

Scenario preparation and review

The first step in preparation of scenarios is elaboration of a table in which all possible future outcomes or alternative future states are listed for each critical issue. When the degree of uncertainty is low or the issue is highly polarized around a very structured set of stable trends, only a couple of future states are identified. For highly complex and uncertain issues, many future states may appear. The table is then analyzed in order to find consistencies among future states of all critical issues. For example, a future state of reduced participation by farmers in the critical issue of research planning and technology diffusion is more consistent with a future state of exploitative use of forest, water, and land resources than with a state of environmental preservation through intense use of modern technologies for sustainable use of natural resources. For complex issues, some of the future states will be unrelated to other critical issues. Occasionally, a given future state of a critical issue will be tied to several future states of other issues. Figure 1 shows how the different states of the critical issues can be combined into scenarios.

The relevance tree approach is useful to this analysis. Each of the "trees" as presented in figure 1 should represent a different unifying theme that underlies the basic structure of an alternative scenario. The resulting scenarios must also be relevant and different with respect to the planning questions addressed.

The next step is to write out each scenario as text, presenting the various hypotheses as facts, explaining the interrelationships identified during the analysis of consistency. Scenario writing is an iterative process of writing, review, additional research, reformulation, and rewriting to obtain distinctly characteristic

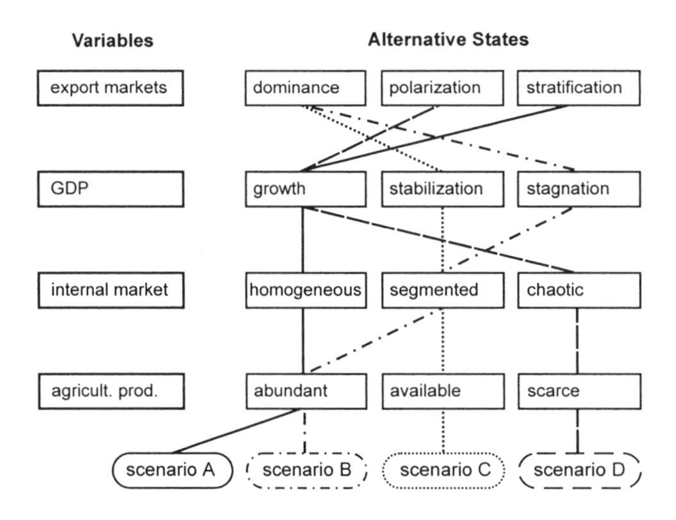

Figure 1. Scenarios in agricultural research in Brazil

Source: Vieira 1999.

themes that integrate the future states of critical issues in a coherent, readable, and plausible manner.

Interpretation and use of alternative scenarios

The interpretation and use of scenarios is highly dependent upon purpose and scope. In policy development, scenarios tend to be value-based and normative in nature, exploring the profiles of impacts on interest groups as defined by alternative actions. This information is valuable in that it contributes to a better definition of possible objectives and articulates priority programs with clearly defined responsibilities among the different agents in a sector. For institutional strategizing and master planning, alternative scenarios should lead to identification of competencies that are required to preserve flexibility in an uncertain context. In program planning, priority rankings of technical objectives may be established given contextual constraints and conditions.

In all cases, the essential contribution of alternative scenarios is to broaden understanding of the relationships between the external context with its inherent uncertainties and traditional planning elements, such as defining objectives, evaluation, choosing actions, and allocating resources. Greater comprehension is especially needed when making decisions in a context of rapid change or uncertainty and in conditions of high complexity. In such cases, it becomes mandatory to rethink basic premises, structures, strategies, and priorities during planning and decision making.

Relevance for agricultural research

Although alternative scenarios are accepted as a valuable approach in the corporate business world and in many national planning offices, they are seldom used in planning agricultural research. Nonetheless, applications have been conducted in Brazil by both federal and state agricultural research organizations and in cooperative efforts with the private sector. A review of these experiences as well as those in other sectors yields a yet incomplete assessment of the relevance and usefulness of alternative scenarios for agricultural research planning.

The reductionism typical of the scientific method yields a limited understanding among researchers of systemic yet subtle change in the agricultural sector. As a result, agricultural research, as other research, tends to be conservative and resist change. Scenarios provide a much needed instrument for counteracting pervasive conservatism in agricultural research planning.

Given the need for a holistic approach in developing scenarios, their preparation requires a multidisciplinary team and intense interactions with all stakeholders. This interaction helps to strengthen linkages between researchers and other agents of agroindustrial complexes.

Although the principles are quite clear, techniques for scenario building are poorly defined, making them difficult to use for agricultural research planning or other applications. Each planning problem requires elaboration of a specific methodology for creating appropriate scenarios, resulting in the need for specialized assistance during the learning period.

Scenario building is labor and interaction intensive and based on an open-ended methodology comprised of art as well as science. Use of this approach must therefore be limited to planning problems characterized by such uncertainty and complexity that one decision or another may lead to completely different outcomes for the organizations involved.

Scenario planning may provide improved insights into the future as well a platform for discussing the best possible plans. This is especially useful when research organizations face major institutional change, such as privatization, mergers, or decentralization, or when the agricultural sector is starting to operate under a new set of parameters, for example, because of trade liberalization, access to a trading block, or a major change in infrastructure policies or regional development goals.

Examples

Two experiences of the Brazilian Corporation for Agricultural Research "Embrapa" are presented here as examples. One relates to institutional planning and the other to research program planning.

In l989 Brazil was at the end of the so-called "lost decade" of economic stagnation and a year before the first popularly elected president in 25 years. Embrapa was in decline, as were most public agencies in the country after years of economic and political turmoil. An multidisciplinary team of 30 researchers devoted six months to learn how to apply alternative scenarios in the question of institutional strategy for the coming decade. After six months of part-time effort, they published a report containing four scenarios and an analysis that called for substantial repositioning of the organization. The report also recommended a broad strategic planning exercise.

The report mobilized latent dissatisfaction within the organization, and the newly appointed management team embraced the recommendations, embarking on an effort to rethink agricultural research in Brazil and to reposition Embrapa in terms of strategy, structure, and process. The effort was extended to Embrapa's 40 research centers, organized around products, ecoregions, service, and thematic issues. The size and diversity of the country and of Embrapa itself caused problems in maintaining coherency of the planning effort. These difficulties were partially offset by the initial scenario report, which provided common language and premises for strategic analysis, as well as giving a holistic and well argued justification for the need for institutional repositioning. Most importantly, the scenarios focused attention on profound but little understood changes that were just beginning to occur in government, agricultural production, and in the market and which could have major implications for the links between research, the agricultural sector, and society in general. After 1993 Embrapa also provided support to state agricultural research organizations, which had requested technical assistance to replicate the process at their level.

Scenario planning was also applied by Embrapa's Grapes and Wine Research Center. In 1991 the center organized a consortium of growers, vintners, suppliers, cooperatives, municipal governments, state agencies, and others in order to establish sector strategies and research program priorities. The study was meant to show the consortium how to respond adequately to the establishment of the Southern Cone Common Market "MERCOSUR." Although little understood, MERCOSUR was considered to be a vital threat to the Brazilian grapes and wine sector, as it would remove barriers to competition from Argentina, where grapes and wine production was low cost and highly productive. A study of the agroindustrial production chain from inputs to final markets was integrated into alternative scenarios describing MERCOSUR trading rules as well as expected economic and market conditions. The study concluded that potential threats were concentrated largely in the popular *vin ordinaire* market segment, but that new export opportunities would open in other segments, especially for grape juice. A sector strategy was developed to exploit these opportunities. The impact on research was significant, in that priority was shifted from developing types of wine towards designing a technological infrastructure for the sector and generating new technology for high-cost inputs such as packaging. The conclusion was completely unexpected by management and staff; it clearly would not have emerged or been accepted had it not been for the cooperative effort in developing and analyzing alternative scenarios for the sector.

References and recommended reading

Candotti, E. et al. 1992. EMBRAPA Global Evaluation Workshop: Report of the External Mission. Secretariat of Strategic Management. Brasília: Empresa Brasileira de Pesquisa Agropecuária.

Castro, C. E. F. and O. C. Bataglia. 1997. Fatores Críticos e Senários para a Pesquisa Agropecuária de São Paulo. São Paulo: Secretaria de Agricultura e Abastecimento de São Paulo, Coordenadoria de Pesquisa Agropecuária.

Embrapa. 1990. Cenários Para a Pesquisa Agropecuária: Aspectos Teóricos e Eplicação na Embrapa. Secretaria de Adminstração Estratégica. Brasília: Empresa Brasileira de Pesquisa Agropecuária.

Embrapa. 1992. II Embrapa Diretive Plan: 1993–1997 (preliminary version). Secretariat of Strategic Management. Brasília: Empresa Brasileira de Pesquisa Agropecuária.

Embrapa 1997. Anais do Seminário Nacional sobre Prospecção Tecnológica. Conselho Nacional de Desenvolvimento Científico e Tecnológico. Brasília: Empresa Brasileira de Pesquisa Agropecuária.

Godet, M. 1987. Scenarios and Strategic Management. London: Butterworths Scientific, Ltd.

Goedert, W., M. L. A. Paez, A. M. G. Castro (eds.). 1994. Gestão em Ciência e Tecnologia: Pesquisa Agropecuária. Brasília: Empresa Brasileira de Pesquisa Agropecuária.

Johnson, B. et al. 1991. Strategic planning in an agriculture research institution with decentralized structure. In *Proceedings of the XVI National Symposium of R&D Management Research.* Brasília: Empresa Brasileira de Pesquisa Agropecuária.

Johnson, B. and J. Marcovitch. 1994. Uses and applications of technology futures in national development: The Brazilian experience. *Technological Forecasting and Social Change*, 45, 1–30.

Johnson, B., M. L. A. Paez, A. Freitas F., and J. B. Araújo. 1991. Alternative scenarios for strategic planning of Embrapa. In *Proceedings of the XVI National Symposium of R&D Management Research.* Brasília: Empresa Brasileira de Pesquisa Agropecuária.

Schoemaker, P. J. H. 1995. Scenario planning: A tool for strategic planning. *Sloan Management Review*, 36 (2): 25–40.

Vieira, L. F. 1999. El Método de Escenarios para Definir el Rol de los INIAs en la Investigacion Agroindustrial. The Hague: International Service for National Agricultural Research.

Waack, P. 1985. Scenarios: Uncharted waters ahead. *Harvard Business Review,* September/October: 72–89.

Wright, J. T. C., S. A. Santos, and B. B. Johnson. 1992. Análise Prospectiva da Vitivinicultura Brasileira: Questões Críticas, Cenários para o Ano 2000 e Objetivos Setoriais. Bento Gonçalves: Empresa Brasileira de Pesquisa Agropecuária, Centro Nacional de Pesquisa em Uva e Vinho.

Chapter 29
Simulation Models for Planning

Philip Thornton

Simulation models have a role to play in the planning of agricultural research, particularly in areas of future work where impacts can be expected to be high and required investments are substantial. Simulation models are useful for communicating the results of alternative research strategies to decision makers and stakeholders. Advances in tools and techniques make simulation modeling increasingly feasible for many organizations. Moreover, there are increasing numbers of well documented and tested models. However, specialized skills are still needed to integrate simulation models effectively into the planning process, especially skills in modeling and communication and facilitating interaction between modelers and planners. If the right skills are available, relatively simple simulation models can offer insights and assess "what if" questions that may be difficult to address using other planning methods.

What is simulation modeling?

Simulation models are abstract representations of particular facets of reality that are built for specific purposes. A simulation model is computer based and, essentially, mathematical. Complete flexibility is allowed as to its underlying structure (unlike a linear programming model, for example). A good simulation model operates on input data to produce output data by mimicking particular processes and parts of reality that are of interest to users. It is typically built for prediction, and this predictive ability can be used for many purposes, one of which is to increase understanding of possible benefits arising from particular research activities.

The range of simulation models is large and growing. This chapter distinguishes simulation models from mathematical programming models, statistical models, econometric models, and spatial models based on geographic information systems, although these (and other) types of models are certainly sometimes referred as "simulation models" in the literature.

Simulation models in agriculture date from the late 1960s. In those early days, they were generally written in high-level programming languages such as FORTRAN. Nowadays they still may be, but increasingly simulation models are being built in commercial spreadsheets and special-purpose dynamic modeling environments. Flexibility is the hallmark of simulation models; there are few rules as to structure or content. Indeed, simulation model building is not an exact science, and most people who have been involved with it stress that art as well as science is involved.

A useful distinction is between "stochastic" and "deterministic" simulation models. This refers to the presence or absence of random (or, more properly, pseudo-random) elements within the model. Stochasticity in a simulation model can be of considerable value, as it allows generation of probability distributions, such as for annual crop yield in response to different weather sequences. Risk can thus be addressed explicitly within the stochastic model framework. Given the biophysical and economic variability associated with agricultural production, it is not surprising that stochastic simulation models are often used to study agricultural systems.

Developing or acquiring simulation models

The effort required to develop simulation models depends on the system to be modeled and the problem to be solved; it may take a few minutes or many years of effort. The usual time involved is somewhere in-between. A problem suitable for study using simulation models has the following features:

- it is basically not an optimization problem (although simulation models are sometimes used for optimization)
- the system under study involves highly dynamic relationships (the time element is important), possibly over many time periods
- the system contains many subsystems that cannot be easily controlled and studied simultaneously
- it is felt that experimentation with the real system is neither feasible nor desirable (such as a country's economy)

But even if a problem seems appropriate for application of simulation models, model building can be expensive and time consuming, and data shortages may mean that simplifying assumptions have to be made that seriously compromise the model's integrity and validity. Recent developments in computer software have yielded some powerful platforms such as Stella and SB Model Maker for constructing dynamic simulation models relatively quickly. Nowadays, a considerable number of simulation models is available off-the-shelf These can be used for some purposes with minimal modification.

Figure 1 shows the classic steps involved in model building (Dent and Blackie 1979):

- *Step 1.* Define the objectives of the modeling, the problem to be addressed, the system to be described, and the level of detail and resolution needed.

- *Step 2.* Collate available data relevant to the problem at hand. If there are gaps, investigate whether these can be filled relatively easily. If not, perhaps the problem or system can be redefined to relax the data constraint.
- *Step 3.* Before starting model construction, assess again whether a simulation model is needed to address the problem. If the model's perceived value outweighs its costs, start model construction.
- *Step 4.* Test the model for internal consistency and for its ability to respond to changes in inputs in a meaningful way (with regard to the objectives of the exercise).

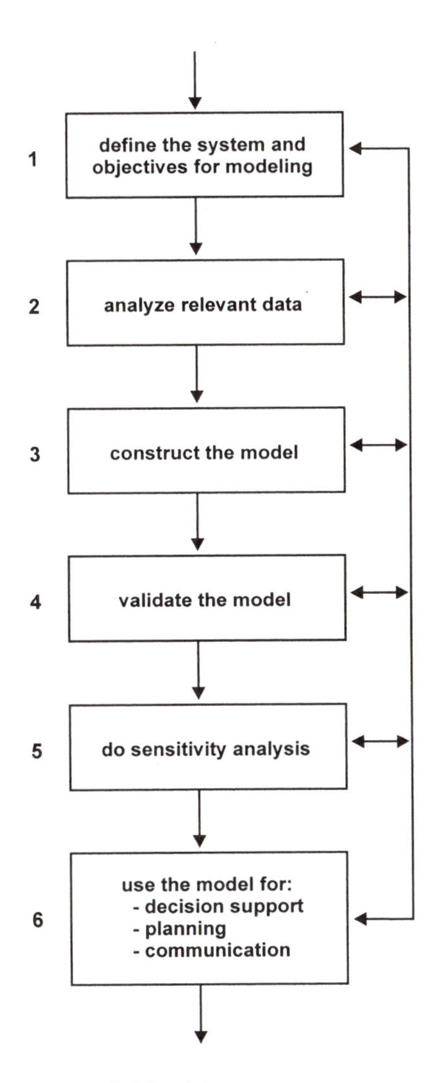

Figure 1. Classic steps in model building

- *Step 5.* Perform sensitivity tests on the model. This is to check the appropriateness of levels of detail in the model and to help define the limits of the model's applicability.
- *Step 6.* Apply the model to the problem defined through simulation experiments, analysis of model outputs, or scenario analysis.

In practice, simulation modeling is a highly iterative and nonlinear procedure that involves combinations of skills that are neither particularly common nor easy to teach (this is the artistic aspect). Unless the model is relatively simple, it is unlikely that simulation models would be developed specifically for agricultural research planning. It is generally much more cost effective to use existing models developed for other purposes, assuming that model and data ownership issues can be resolved.

Integrating simulation models into planning

Many simulation models are constructed as outputs of research activity. That is, they are designed as frameworks or statements about the state of knowledge of particular sets of processes at a particular time. Crop growth and development simulation models are good examples of this. Such models tend to be highly detailed and complex, and they are unlikely to be constructed for research planning per se. But if they do exist, they can be a valuable tool for identifying areas where knowledge is weak and effort and research money might usefully be expended to improve understanding.

Increasingly, simulation models are constructed more as inputs to the research process itself, in the form of conceptual models that are built either from first principles or from in-depth understanding of the current state of knowledge. Such models are often designed to generate information on the feasibility of particular lines of inquiry (including the development of detailed, complex models of the processes under study). They tend to be relatively simple and easy to understand, but usually lack the details and refinement necessary to address the management of agricultural systems. In research planning, these types of models have a role to play in identifying possible lines of attack on complex problems and, sometimes, in detecting altogether new areas of research.

Another important role of simulation models, especially with recent advances in user-friendly software, is as a tool for communication. In this mode, models can illustrate, sometimes graphically, the consequences of different courses of action (e.g., different management techniques or outcomes of various research initiatives) to agricultural research stakeholders. This might be aimed at persuading donors and policymakers of the merits of a particular line of research or at resolving conflicts between stakeholders with radically different attitudes and objectives.

Using these methods, input can be garnered for agricultural research planning at particular times. Similarly, planners should not forget that planning information can often be used in or generated by simulation models in conjunction with other types of model and other analytical frameworks. One fairly common multistep analysis is to use a simulation model to generate distributions of output in response to particular inputs, and then to use these input-output relationships in a mathematical programming model to find ways to optimize resource use. Such an analysis was done for land-use planning in Costa Rica (Jansen et al. 1997) and Brazil (Veloso et al. 1994).

Relevance for agricultural research

The flexibility of simulation models means that their scope is very wide. Models may be constructed to answer analytically challenging questions concerning complex systems. For example, what are the benefits and trade-offs involved in feeding maize residue to livestock compared with returning it to the land to supply nutrients to subsequent annual crops? Models may also be constructed to support decision making and to illustrate the possible consequences of a decision to stakeholders. For example, what are the likely impacts at household and regional levels of particular off-take rates of water in a river-based irrigation scheme?

The benefits of a validated simulation model stem from its ability to generate experimental results without actually having to do the experiments and from the ability to quantify the costs, benefits, risk, and variability inherent in particular systems and courses of action. Simulation models can offer insights into radically new areas of research for an organization. In cases where there is little concrete information to go on, a relatively uncomplicated simulation model can show possible economic and biophysical outcomes of particular interventions in components of agricultural systems. Even at a superficial level of detail, such information can be useful for priority setting and research planning. In fact, it may be difficult to generate in any other way.

In addition to the generally high cost (particularly in terms of time) of simulation modeling, there are a number of pitfalls. Perhaps the most common is that modelers get so involved with their models that they get tied up in details, losing sight of the goal of the exercise. There is a fine trade-off between the cost and effort to develop a model and the value of the information produced. Ability to stand back and assess this objectively is extremely important. A related pitfall is the common perception that simulation models are merely black boxes, with inner workings hidden from sight. This can seriously affect credibility of simulation results and institutional uptake of ideas that might be generated. It may be possible to side step this issue by using simpler models or existing models with long track records in the field, or by educating personnel in how the models work. Another pitfall (or perceptional problem, at least) is the idea of simulation modeling as a lone pursuit. Research planning is a group process with many people and many steps. If simulation models are to be integrated into the process, it must be done carefully. Really, the role of simulation models is limited to providing specific information at specific steps in the process.

Weighing the costs and benefits of using simulation models in research planning is unlikely to be easy. But if the skills exist within the organization under consideration or can be borrowed easily, if the problem domain is complex, and if radical shifts are envisaged in the portfolio of research activities, simulation modeling should be seriously considered.

Examples

Builders and users of simulation models often pay lip service to the notion of generating information for research planning. But few good examples exist of effective utilization of simulation models in the planning process. It is probably fair to conclude that to date, the contribution of simulation models to agricultural research planning is minor but increasing. Their potential remains to be exploited. This section illustrates the use of simulation models in agricultural research planning in response to five broad questions that may be asked during the planning process. Some of the studies highlighted were designed explicitly for agricultural research planning, while in others, subtle insights into subsequent or potential research activity was almost a by-product of the effort.

Is a particular piece of research worth doing, assuming it could be successful and adopted?

The International Livestock Research Institute (ILRI) recently completed an impact assessment that measures the potential returns to trypanosomosis vaccine research. The activity involved measuring the potential productivity impact of successful trypanosomosis control using a herd simulation model; linking model results to spatial databases to determine where the potential increase in livestock productivity from a new vaccine was likely to occur; and valuing economic returns to this area of research, given various assumptions about probability of research success and adoption levels.

A critical step in the analysis was to estimate the productivity of cattle herds before and after control of the tsetse fly, the vector of the disease. Data on livestock productivity and herd structures from field studies were used as inputs to a 10-year herd simulation model. The model predicted annual milk and meat production for the herd both before and after disease control technology (tsetse control) was introduced. Outputs from the model, including production figures, were then valued using an economic surplus model and extrapolated to other parts of Africa where livestock populations are at risk. The potential productivity gains imply that a vaccine could result in significant reduction in the cost of producing milk and meat for African farmers, leading to increased meat and milk supply worth over US $300 million in lower consumer prices. The net present value (the discounted benefits minus the discounted costs) of ILRI and collaborators' trypanosomosis vaccine research was estimated at $118 million, with an internal rate of return (one measure of the return to investment) of 25 percent. Thus, even though the search for a vaccine is long and difficult, these figures help to maintain and justify the major research effort in a climate of declining research funding (Kristjanson et al. 1998).

Of all available options, which offers the highest potential gains?

A study of research priorities for stimulating oyster production was undertaken in Virginia, USA. A relatively simple bioeconomic simulation model was constructed to simulate oyster production under uncertainty including growth, disease, and economic components. This model was used to discern what effects different types of research information would have on economic returns to a representative oyster production enterprise. From a wide range of options studied, the model showed that profitability of oyster production would be increased most by improved seed-harvest technologies and accurate knowledge of the salinity threshold at which mortalities occur due to disease. The simulation model was built specifically for the study (Bosch and Shabman 1990).

Other studies have used existing simulation models as well. An example is the use of an animal production model built originally in the USA, modified in the UK, then modified again for Latin America, to assess methods of using planted, improved pasture in the extensive grazing systems of Latin America's acid savannas. The model simulated animal growth, development, and production in response to particular diets, aggregated to the herd level. Economic evaluation of different 10-year scenarios was then undertaken to assess promising alternative management strategies. One of the most severe constraints to production system profitability, insufficient milk availability for sale, was identified for subsequent testing in the field research program of the institute involved (Thornton 1989).

What is the status of understanding of a set of particular processes?

In many countries of sub-Saharan Africa the "yield gap" is a cause of concern to researchers, development agencies, and governments. The yield gap refers to the difference between the genetic yield potential of a crop (often estimated by what is achieved on experiment station plots with high levels of inputs) and the yield commonly achieved in farmers' fields. Farmers' yields of maize in many parts of Africa are on the order of 1.0 to 1.5 tons per hectare, while modern hybrids (with fertilizer) easily yield 7.0 tons per hectare. There are many reasons for this. Farm households might prefer the taste and cooking characteristics of local maize varieties, and yields of these local varieties are often more stable than those of hybrids under low input and uncertain rainfall conditions, to name just two. There are also biophysical reasons, which can be investigated using a tool such as the crop-growth simulation model "CERES-Maize."

This model requires a minimum set of soil, weather, genetic, and management information. Researchers from various institutions in Kenya did studies using field data from 70 sites in the country's maize-growing regions. That showed various factors as contributing to the yield gap for maize in Kenya: phosphorus and nitrogen deficiencies, the parasitic *striga* weed, and water-logging in certain soils, in particular. The model is sensitive to nitrogen, and phosphorus routines are being developed, but the model has no sensitivity to water-logging (water-logging occurs when the water table rises to within a few centimeters of the soil surface, killing the roots of the plant and thus restricting its nutrient and water uptake). Accordingly, research planners decided to commission the construction of routines in CERES-Maize that can simulate the effects of water-logging on maize growth and development. Field trials were carried out in South Africa to generate the data required.

With this data, modifications were made to the components of CERES-Maize that simulate soil water drainage and to calculate an oxygen stress index for crop growth and development. With these modifications, CERES-Maize is being used to quantify the identified components of the yield gap. It appears that water-logging accounts for about 100 kilograms per hectare of "lost" yield (on average, across all soil types in maize-growing zones) in Kenya every season. The ability to partition yield components and yield loss in this way has important implications for planners, as it allows costs and benefits of various research interventions to be quantified reasonably objectively (Du Toit 1998).

Which crop characteristics are most advantageous for a particular environment?

Various crop simulation models can predict growth and development as affected by soil and weather conditions, agronomic practices, and cultivar traits. These have been used to examine the effects on yield of specific traits that represent possible ideotypes. By using data from a number of seasons, variability arising from unpredictable water deficit or other weather effects can be quantified. Such models are suitable for helping breeders understand genotype by environment interactions. They can provide an independent estimate of site productivity and allow assessment of specific environmental factors. By examining responses to temperature, photoperiod, soil moisture, and nutrients, scientists can determine the mechanisms of adaptation or critical factors determining crop response. This information, in turn, can lead to more reliable characterization of growing environments and grouping of cultivars.

Understanding problems of adaptation can be as simple as using a simulation model to examine water balances for different seasons or locations. In studying the contrasting responses of different bean cultivars in the Mexican highlands, researchers used "BEANGRO," a detailed simulation model of the growth and development of *phaseolus* bean, to examine the length of growing season at three sites over many years. For two sites, long growing seasons and early onset of the season were associated with a greater probability of adequate rainfall. At the third site, total rainfall was lower and uncorrelated with onset or length of season. A cultivar with a growth cycle that lengthens with early planting would suit the first two sites, while a cultivar with a constant, short growing cycle would be better suited for the third site. Such analyses are leading research planners and scientists to examine the most effective ways in which varieties can be tested for maximum benefit (Acosta and White 1995, White 1998).

How might this research area be reoriented?

East Coast fever is another livestock disease for which ILRI is working to develop a vaccine. East Coast fever causes high mortality in susceptible cattle and is found throughout southern and east Africa. The life cycle of the protozoan parasite that causes East Coast fever, *Theileria parva*, is extremely complex; and the parasite is injected into cattle by infected ticks of the family *Rhipicephalus*. The parasites that infect cattle appear to have originated in parasites that occurred in buffalo, and buffalo remain a source of infection for indigenous tick populations. Research on immunization strategies at ILRI and elsewhere concentrate on the bovine host's immune response to two particular stages in parasite's life cycle. An effective vaccine against East Coast fever will probably include a number of components that induce different immune responses to infection.

A major consideration in field use of such a vaccine is the way in which it might influence the epidemiological state of the disease. East Coast fever exists in a number of epidemiological forms, depending largely on the density of populations of infected ticks, which itself varies according to climate, cattle management, and tick control measures. How sustainable would such a vaccine be in the field under different management conditions, in a situation of high endemic stability (i.e., high density of ticks along with high animal antibody prevalence, so as to result in low disease incidence and low case fatality)? To help answer this question, a model of the variability within *T. parka* strains was built that predicts the ability of different cattle to generate a protective response against the tick challenge.

The model itself is very simple and can be coded in half a page of FORTRAN, although the insight and background knowledge associated with it was enormous. The major implications of the model are that no single strain of the parasite has broad immunization potential in cattle, and that because certain animals respond ineffectively to the parasite, the parasite could well exert selective pressure on the cattle population. In addition, the model predicts that incorporation of two or more components in a vaccine will substantially enhance its protective effect.

Prospects for a vaccine to seriously affect the endemic stability of the disease in particular management systems is having substantial impact on the development of ILRI's East Coast fever vaccine strategies. Such information helps both planners and researchers to consider how best to plan and evaluate vaccines. Model results underline very clearly that vaccine evaluation must take place under different epidemiological circumstances (different locations and season types). This has obvious implications for the costs associated with, and length of time necessary for, this research (McKeever and Morrison 1998).

Without implicating them in any way, I thank Patti Kristjanson, Declan McKeever, and Bill Thorpe for comments on a previous draft of parts or all of this chapter.

References

Acosta-Gallegos, J. A. and J. W. White. 1995. Phenological plasticity as an adaptation by common bean to rainfed environments. *Crop Science,* 35 (1): 199–204.

Bosch, D. J. and L. A. Shabman. 1990. Simulation modeling to set priorities for research on oyster production. *American Journal of Agricultural Economics,* 72 (2): 371–381.

Du Toit, A. 1998. The Development of a Simple Water Logging Subroutine for Water-Table Soils. Research Report. Potchefstroom, South Africa: Agricultural Research Centre.

Jansen, D. M., J. J. Stoorvogel, and J. G. P. Jansen. 1997. A quantitative tool for regional land-use planning. In *Applications of Systems Approaches at the Farm and Regional Levels* edited by P. S. Teng et al., pp. 399–411. Dordrecht: Kluwer Academic Publishers.

Kristjanson, P. M., B. M. Swallow, G. J. Rowlands, R. L. Kruska R L, and P. N. de Leeuw. 1998. Measuring the costs of African animal trypanosomosis, the potential benefits to its control, and the returns to research. *Agricultural Systems* (submitted).

McKeever, D. J. and W. I. Morrison. 1998. Novel vaccines against *Theileria parva*: Prospects for sustainability. *International Journal for Parasitology,* 28 (5): 693–706.

Thornton, P. K. 1989. Computer experimentation with an energy-based simulation model of animal production in the eastern savannas of Colombia. *Tropical Agriculture,* 66: 217–220.

Veloso, R. F., M. J. McGregor, J. B. Dent, and P. K. Thornton. 1994. Técnicas de modelagem de sistemas aplicadas em planejamento agrícola dos cerrados. *Pesq. Agropec. Bras., Brasília,* 29 (12): 1877–1887.

White, J. W. 1998. Modeling and crop improvement. In *Understanding Options for Agricultural Production* edited by G. Y. Tsuji, G. Hoogenboom, and P. K. Thornton: 179–188. Dordrecht: Kluwer Academic Publishers.

Further information

Important books on simulation models, in chronological order

Shannon, R. E. 1975. Simulation: The Art and Science. Englewood Cliffs: Prentice-Hall.
Fairly non-technical account, but somewhat dated now, of simulation, with examples of applications in various fields other than agriculture.

Dent, J. B. and M. J. Blackie. 1979. Systems Simulation in Agriculture. London: Applied Science Publishers.
A good, brief introduction to simulation models in agriculture, although it would benefit from updating.

France, J. and J. H. M. Thornley. 1984. Mathematical Models in Agriculture. London: Butterworths.
Rather more technical than Shannon or Dent and Blackie, but a good source of information on important plant and animal processes and how these can be modeled.

Kleijnen, J. and W. van Groenendaal. 1992. Simulation: A Statistical Perspective. Chichester: John Wiley & Sons.
Not as good as a general introduction to simulation, but strong on Monte Carlo and other statistical aspects, with examples of economic and corporate models and operations research.

Leffelaar, P. A. (ed.). 1993. On Systems Analysis and Simulation of Ecological Processes, with Examples in CSMP and FORTRAN. Dordrecht: Kluwer Academic Publishers.
Technical account of models of plant processes, of more interest to modelers than to planners.

Tsuji, G. Y., G. Hoogenboom, and K. K. Thornton (eds.). 1998. Understanding Options for Agricultural Production. Dordrecht: Kluwer Academic Publishers.
Account of a 10-year project funded by the US Agency for International Development; contains highly technical chapters on model workings, but other less technical chapters on application of crop models to a wide range of problems.

Websites

CAMASE: "Concerted Action for the development and testing of quantitative Methods for research on Agricultural Systems and the Environment"
http://www.bib.wau.nl/camase
The CAMASE register of agroecosystem models contains simulation or optimization models that are documented, at least at a scientific level, and "validated," at least partially. It also describes software tools that are directly related to the simulation activities. The register is at http://www.agnic.nal.usda.gov/agdb/camram.html.

DSSAT: "Decision Support System for Agrotechnology Transfer"
http://everex.ibsnat.hawaii.edu

DSSAT is an example of computer software that combines crop, soil, and weather databases and programs to manage them, with crop models and application programs, to simulate multi-year outcomes of crop management strategies. DSSAT allows users to ask "what if" questions and simulate results by conducting field experiments that would otherwise consume a significant part of an agronomist's career. DSSAT has a listserver with about 300 subscribers.

ICASA: "International Consortium for Agricultural Systems Applications"
http://agrss.sherman.hawaii.edu/icasa
ICASA brings together systems scientists associated with DSSAT (see above) in a unified effort to meet potential interests of the international and national programs in systems approaches to problem solving. ICASA is involved with a range of systems tools, technical support services, and training courses and programs.

Glossary

Gerdien W. Meijerink

Planners need to speak the same language – that is, agree on the meaning of the words and concepts they use. Yet planning terminology differs greatly from one country to another, between organizations, and even within a single organization! To complicate matters further, partners in research program planning often come from different places and have different disciplinary backgrounds. It is less important which terms and definitions are used, as long as members of a single planning group share a common vocabulary.

This glossary presents definitions of many terms used in this sourcebook in alphabetical order. In some cases different definitions are provided to reflect differences in the professional literature.

Accountability: The obligation of an organization or its members to account for, report on, or explain their actions and the use of resources entrusted to them. Originally, accountability referred mainly to compliance with established norms of financial management. In recent years, the meaning of accountability has broadened to include the achievement of performance targets and compliance with norms external to the organization – such as protection of human rights or preservation of the environment.

Activity: 1. An element of work performed during the course of a project. Activities have an expected duration and resource requirements. They are often subdivided into tasks (Project Management Institute 1996). 2. The smallest self-contained unit of work used to define the logic of a project. Activities have a definite duration, and logical relationships to other activities in a project, use resources such as people, materials or facilities, and have an associated cost (Welcom Glossary). 3. A component of a research project. A research activity is a coherent set of specific actions to be carried out in a given period. All activities are necessary to attain the desired result. It is generally monodisciplinary and carried out by one researcher, sometimes with the assistance of technicians, according to an experiment protocol. It has an objective, location, schedule, and cost. A research activity can have several locations if the desired objective applies to several agroecological zones. The number of activities varies from project to project (Kissi n.d.).

Agricultural research policy: A framework guiding investments and activities in the generation and dissemination of agricultural knowledge and technology in a country. It is developed in a process that links innovation in agriculture to prospects for growth and development in the agricultural sector and in the broader national economy.

Annual planning: A process that has direct operational implications and that is derived from a medium-term plan (where such a plan exists). An annual plan specifies the steps in implementation for a given annual budget cycle to which it is directly linked. As implementation progresses, unanticipated changes in objectives, programs, and resource availability require a feedback link to revise both the annual and the medium-term plan. This is a part of the task of developing a rolling plan process. What are called annual plans are in fact exercises in program planning and budgeting (Retzlaff 1992).

Assumptions: 1. Something that is taken for granted or advanced as fact (Webster). 2. In the logical framework, externally determined factors that are decisive for the success of the project or activity but that are beyond the control of the project's or activity's mandate area (Kronen 1996). 3. In the logical framework, important assumptions are key threats to the project that exist in the external environment. The process by which teams specify assumptions resembles scanning for external threats in strategic planning. The process helps teams to anticipate where a project may fail and develop strategies to protect against threats. If the project is based on heroic or "killer" assumptions, then it may not be viable (Sartorius 1996).

Beneficiaries: A particular person or group who has benefited from a development activity. Who exactly are beneficiaries is evident only in hindsight. Intended beneficiaries may not have benefited after all. Therefore it may be clearer to refer to the people who are the intended recipients of benefits as "participants."

Benefits: 1. The enhanced efficiency, economy, and effectiveness of future research operations to be delivered by a program (Welcom Glossary). 2. Valued outcomes or processes (Scriven 1991). 3. Comparing different benefits can be difficult. Therefore, it can be useful to distinguish benefit dimensions. For instance, benefits in agricultural research may be gained in five dimensions: cropping, livestock, forestry, soil quality, and water management. In priority-setting exercises, activities, projects or programs can be scored against these dimensions, avoiding double counting (Janssen and Kissi 1997).

Budget: 1. The various cost estimates for project funding or a list of identified and expected annual revenues and expenses to be executed. The budget is a critical tool in translating plans into research actions. Budgeting is translating operational short-term agricultural research plans into financial terms, so that limited available financial resources can be applied in the most efficient manner to carry out the agricultural research activities described in that plan. 2. The planned cost for an activity or project (Welcom Glossary).

Committee: See *team*.

Constraint: 1. A situation or factor that prevents production potential from being fully achieved. This potential may be based on extending the area under cultivation, increasing yields, cutting production costs and losses, or raising value added by processing and packaging (Janssen and Kissi 1997). 2. Applicable restrictions that will affect the scope of the project (Welcom Glossary).

Constraint tree: 1. A hierarchy of research and development problems originating from the major constraint. It is a tool for systematically analyzing such problems, allowing for a participatory approach to formulating research projects and programs. The starting point of building a constraint tree to identify the central constraint. The causes and effects of all factors underlying the constraint are then analyzed, along with their interdependency. The constraint tree takes the form of a flow chart composed of various boxes at different levels.

Contingency planning: The development of a management plan that uses alternative strategies to ensure project success if specified risk events occur (Welcom Glossary).

Cost-benefit analysis: 1. The analysis of the potential costs and benefits of a project to allow comparison of the returns from alternative forms of investment (Welcom Glossary). 2. Monetary method to weigh costs and benefits through time by using a discount factor. The higher the discount factor, the more weight is placed on costs and benefits that occur in the near future. The cost-benefit analysis (or CBA) usually results in a net present value (NPV) or internal rate of return (IRR). 3. A CBA estimates the overall cost and benefit of each alternative (product or program) in terms of a single quantity, usually money. This analysis will, where feasible, provides an answer to the question, "Is this program or product worth its cost?" Or, "Which of the options has the highest benefit/cost ratio?" This is only possible when all the values involved can be converted into monetary terms. This is normally not possible in the case of ethical, intrinsic, temporal, or aesthetic elements (Scriven 1991).

Delphi method: A technique to arrive at a group position regarding an issue under investigation, the Delphi method consists of a series of repeated interrogations, usually by means of questionnaires, of a group of individuals whose opinions or judgments are of interest. After the initial interrogation of each individual, each subsequent interrogation is accompanied by information regarding the preceding round of replies, usually presented anonymously. The individual is thus encouraged to reconsider and, if appropriate, to change the previous reply in light of the replies of other members of the group. After two or three rounds, the group position is determined by averaging (Principa Cybernetica Web).

Evaluation: 1. Judging, appraising, or determining worth, value, or quality or research – whether it is proposed, ongoing, or completed – in terms of its relevance, effectiveness, efficiency, and impact (Horton et al. 2000). 2. A process which attempts to determine as systematically and objectively as possible the relevance, effectiveness, efficiency, and impact of activities in the light of specified objectives. It is a learning and action-oriented management tool and organizational process for improving both current activities and future planning, programming, and decision making (UNICEF 1990). 3. Evaluations are analytical assessments addressing results of public policies, organizations, or programs. They emphasize reliability and usefulness of findings. The role of evaluation is to improve information and reduce uncertainty. However, even evaluations based on rigorous methods rely significantly on judgment. A distinction can be made between ex-ante evaluations and ex-post evaluations (OECD).

Experiment: An operation carried out under controlled conditions in order to discover an unknown effect or law, to test or establish a hypothesis, or to illustrate a known law (Webster).

Experiment planning: Experiment planning represents the lowest level of agricultural research planning. It is targeted at identifying the most efficient and effective option to achieve the experiment results necessary to develop a required technology.

Financial planning: Financial planning for research institutes aims to reconcile the level of research activity with the likely availability of funds from different sources. Financial planning requires strategies for: (a) identifying and developing alternative sources of funding; (b) using and allocating funds in the most efficient manner; and (c) adjusting program and institute size to the projected funding base.

Foresight study: The process involved in systematically attempting to look into the longer term future of science, technology, the economy, and society with the aim of identifying the areas of strategic research and the emerging generic technologies likely to yield the greatest economic and social benefits.

Gender analysis: 1. Aims to make explicit social roles and relations. The purpose of gender analysis is to take into account in research and planning other factors and realities that affect, or are influenced by, gender relations. Through this approach it is possible to tailor interventions to meet women's and men's specific gender-based constraints, needs, and opportunities, thereby increasing the effectiveness and efficiency of agricultural research. 2. Seeks answers to fundamental questions such as who does or uses what, how, and why. The purpose of gender analysis is not to create a separate body of social knowledge about women, but to rethink current processes – such as neutral resource use and management, economic adjustment and transformation, or demographic changes – to better understand the gender factors and realities within them.

Geographical information system: (GIS) Helps manage and visualize large amounts of data in agricultural research planning. Biophysical and socioeconomic data, spatially referenced, are stored in a database. The database can be used to draw maps illustrating distribution of production, environmental, marketing or socioeconomic problems. Such maps may help the decision makers to analyze and prioritize research problems. GIS can also be used to define homogenous regions, which can be used to target research efforts or to define similarities between regions and the potential for technology dissemination.

Goal: 1. Ultimate objective to which a research activity, project, or program contributes. It is often outside the domain of research. 2. Higher order objective or longer term impact that a research project seeks to achieve (Sartorius 1996). 3. A one-sentence definition of what will be accomplished incorporating an event signifying completion (Welcom Glossary).

Group: See *team*.

Impact: 1. The broad, long-term significant effects resulting from research (based on Horton et al. 2000). 2. A term indicating whether a project had an effect on its surroundings in terms of technical, economic, sociocultural, institutional, and environmental factors (UNICEF 1990). See also *outcome*, *output*, and *research result*.

Impact assessment: 1. Assessing the pros and cons of pursuing a particular course of action (Welcom Glossary). 2. Evaluation of the extent to which a program or project causes changes in the desired direction in a target population (Henry 1985). 3. Any effect – whether anticipated or unanticipated, and positive or negative – brought about by a development intervention, at the level of individual or the organization (Horton et al. 2000).

Implementation: 1. To give practical effect to, and to ensure actual fulfillment of a research policy, project plan, or program plan by concrete measures (based on Webster). Often implementation receives less attention than the planning process.

Logical framework: (logframe) A tool for planning programs and projects in the broader context of development goals. By leading planners step-by-step through the cause-effect relationships between activities, outputs, and goals, it helps link program inputs and objectives in a clear, logical way. Research managers use the logical framework to link research program and project objectives to national goals. A logical framework captures and documents the collective thinking behind the project and guides subsequent project investment, monitoring, and evaluation.

Long-term research plan: Commonly a document that describes the proposed growth of the system in terms of its research priorities and programs in the next 10–15 years and the human and other resources needed for this purpose. It is more relevant to research systems that have already made significant advances in institution building. The long-term research plan cannot be considered the equivalent of a national agricultural research plan. For one thing, it is not such a comprehensive document and the element of strategy is often lacking. For another, the emphasis here is more on the research program rather than all the system-building factors that have to be stressed in preparing a national agricultural research plan (Jain 1990).

Management: See *project management*.

Management information system: (MIS) 1. An ongoing data collection and analysis system, usually computerized, that allows timely access to service delivery and outcome information (Rossi and Freeman 1985). 2. An MIS is a system using formalized procedures to provide management at all levels with appropriate information, based on data from internal and if desired also external sources, to enable them to make timely and effective decisions for planning, directing, and controlling the activities for which they are responsible.

Master plan: 1. A process by which a national research system analyzes its present and future environment, defines its medium and long-term goals and objectives, and develops a plan based on priorities and available resources to attain these objectives and goals. An integral part of this process is the building of a sustainable research capacity, which would enable the system to respond appropriately to the changing needs of the agricultural industry, and the nation. All these must be based on a clear and realistic vision of the future of research, the design of relevant and effective programs, and an assessment of the resources needed to achieve this vision. 2. The term master plan comes closer to the national agricultural research plan. One might say that it is a popular term for a national agricultural research plan (Jain 1990).

Medium-term plan: 1. National agricultural research plans on a five-yearly basis (Jain 1990). 2. For a period of two to five years, a plan that outlines objectives to be achieved and the activities required. A medium-term plan is more concrete than a strategic plan, but less detailed than an annual plan. It is especially useful for resource planning, for example personnel. 3. Involves five to seven years, and is formulated within the framework of established national development and agricultural research objectives. Where a strategic plan has been prepared, the task of a medium term plan is the further specification of research objectives; the determination of research programs and priorities; an iterative assessment of resources – human, physical, fiscal, and information – for implementation; and the identification of possible constraints and the steps required to relax them (Retzlaff 1992).

Milestones: An activity with zero duration (usually marking the end of a period) (Welcom glossary). 1. Moment in time when intermediate outputs of a project should be available. Milestones are used to organize the planning of long-term projects into smaller time intervals, and to monitor the progress of the project.

Mission: 1. Official statement of the reason for an organization's existence – its basic goals and purpose (Horton et al. 2000). 2. Brief summary, approximately one or two sentences, that sums up the background, purposes, and benefits of the project or organization (based on Welcom Glossary).

Monitoring: 1. Observing or checking on research activities and their context, results, and impact. The goals of monitoring are to ensure that an activity is proceeding according to plan, to provide a record of input use, activities, and results, and to warn any deviation from its initial goals and expected outcomes (Horton et al. 1993). 2. The capture, analysis, and reporting of project performance, usually as compared to plan (Project Management Institute 1996). 3. The periodic oversight of the implementation of an activity which seeks to establish the extent to which input deliveries, work schedules, other required actions and targeted outputs are proceeding according to plan, so that timely action can be taken to correct deficiencies detected. Monitoring is also useful for the systematic checking on a condition or set of conditions, such as following the situation of women and children (UNICEF 1990).

National agricultural research program: The national agricultural research program can be considered at two levels. In a broad sense the national agricultural research program indicates the country's priorities in terms of commodities and natural resources and factors of production, with indication of relative allocation of funding support. It describes the major objectives of research but does not attempt to give the detailed technical content of the different research thrusts proposed. The national agricultural research program of a more detailed kind is a technical document, which lists the research programs, projects and experiments together with resource allocation of all the institutes and experiment stations forming part of the national agricultural research system (Jain 1990).

National agricultural research plan: A strategic document that describes the planned evolution of a national agricultural research system in the next 10–15 years for the best possible use of its resources and opportunities. It must deal with problems of structure and organization and the development and management of its human and other resources, as well as the linkages of the research system. Logically, the preparation of a national agricultural research plan should follow from a comprehensive review of the system as it exists at present. This should help to identify the action needed for building the system, keeping in view the contributions it will be called upon to make (Jain 1990). See also *master plan*.

Objective: 1. Predetermined results toward which effort is directed (Welcom Glossary). 2. An aim that one hopes to achieve. An objective may be derived from a goal but it is more specific than a goal, although not necessarily expressed quantitatively (Van Staveren and Van Dusseldorp 1980). 3. The effect which is expected to be achieved as the result of the project (NORAD 1990).

Organizational performance: 1. Organizational performance can be evaluated by criteria of effectiveness, efficiency and sustainability. The factors that influence performance are an organization's (i) mandates, objectives and policies, (ii) organization, structure and linkages, (iii) resources and information and (iv) program planning and management. 2. Organizational performance is a function of the organization's external environment, motivation and capacity. It can be measured by the effectiveness with which it achieves its mission and goal, the efficiency of resource use and the organization's sustainability in terms of its continued relevance to stakeholders (based on Lusthaus et al. 1995). 3. Organizational performance is the ability to plan and use resources in the production of outputs that are relevant and useful for the organization's target users or clients. A performance oriented agricultural research organization is necessarily focused on producers/users and on research output productivity. This definition of performance highlights two important dimensions, one related to productivity and outputs, and the other related to relevance of these outputs for an organization's stakeholders, an aspect closely linked to the idea of accountability.

Outcomes: When users adopt outputs of research (e.g. new technologies), they can transform these into positive outcomes such as improved production, cost reduction, and profits. Outcomes therefore are less tangible than outputs (Based on Hartwich 1998). See also *impact*, *output*, and *research result*.

Outputs: 1. Outputs are the key components of a research project. They are the results for which the project is held directly accountable and for which is given resources. They are the main deliverables of a research project (Sartorius 1996). 2. The results that the project management should be able to guarantee (NORAD 1990). 3. The direct results of an intervention; a "deliverable" for which management is responsible (Horton et al. 2000). 4. Outputs are more or less measurable products of the research process, they indicate that new and advanced knowledge has been acquired (Hartwich 1998). See also *impact*, *outcome*, and *research result*.

Participatory approach: 1. A participatory approach implies a process (e.g., in planning, evaluation, and research) that involves the clients and stakeholders at different levels and different stages. It is important to recognize that stakeholders can 'participate' to different degrees and with different roles. Participation may range from simple consultation with stakeholders on their wants and needs to more collaborative forms of joint planning, to planning efforts in which end users have real decision-making power in selecting priorities by linking joint planning to joint evaluation and accountability sharing. 2. Participation in agricultural research planning implies that the stakeholders are involved in setting research agendas. They may be involved at different levels of planning (e.g., national, regional, local) and at different stages (e.g., setting broad agricultural goals, developing agroecological strategies and defining local community research priorities).

Performance: See *organizational performance*.

Plan: 1. A plan is a document that lays out a specific set of objectives and the means of accomplishing them (Horton et al. 1993). 2. A plan is an intended future course of action. It is the basis of the project controls (Welcom Glossary). 3. A method devised for making or doing something or attaining an end (Webster). See also *project plan.*

Planning: 1. A process for setting organizational goals and establishing the resources needed to achieve them. It is also a way of building consensus around the mandate, direction, and priorities of a research program or organization (Horton et al. 1993). 2. The process of identifying the means, resources and actions necessary to accomplish an objective (Welcom Glossary).

Planning by objective: A systematic method for planning research activities for a given domain or subsector. Specific intermediate research objectives are determined as a means of achieving a defined overall objective. Intermediate objectives are then translated into research projects composed of research activities, the results of which contribute to the achievement of the research project's specific objective (Kissi n.d.).

Planning group, team, or committee: A temporary analytical, coordinating, and advisory body whose job is to plan, monitor, evaluate, and adjust research activities for a given domain or subsector. It is a structure within which a dialogue is established among researchers and the users of research results for the purposes of program formulation. The researchers come from various disciplines and belong to different research bodies. The users are selected according to the type of research program to be formulated. A more formal and permanent planning group, team or committee can be defined as planning unit.

Planning horizon 1. End date of a plan. Beyond this time limit it is not feasible or possible to do this type of planning. 2. In formulating projects one or more time horizons must be assessed. The time horizon of a long-term plan may cover more than five years. Another time horizon is the 'near future', covering only a few years and the 'medium-term' which covers three to five years (based on Van Staveren and Van Dusseldorp 1980).

Planning tools 1. The means or procedures used in attaining a plan (Webster). 2. Planning tools contribute to the planning process in two dimensions: (a) by supporting the planning process by laying out clear procedures and clear expected outputs and by defining rules for communication, (b) by allowing for the integration within the planning process of external information.

Planning unit: The formal unit responsible for planning formulation procedures, organizing meetings and preparing the necessary background information. This unit ensures that the plan under formulation is consistent with the overall direction of the national agricultural research systems.

Policy: A high-level overall plan embracing the general goals, guiding principles and acceptable procedures especially of a governmental body.

Priority setting: 1. Deciding on the relative importance of research areas or projects, usually in terms of their expected contribution to organizational or development goals (Horton et al. 1993). 2. Broadly defined, agricultural research priority setting is the process of making choices amongst a set of potential research activities (Mills 1998). 3. Priority setting contains several elements: (a) identifying the objectives of research, (b) defining the relevant alternatives to be assessed, (c) assessing the effects of the alternatives and evaluating those effects in relation to the objectives, and based on the evaluation, (d) comparing the alternatives and making selections (based on Alston et al. 1998).

Program: 1. An organized set of research projects, activities, or experiments that are oriented towards the attainment of specific objectives. A program is not time-bound, as projects are, and programs are higher in the research hierarchy than projects. The word program can refer either to the actual set of research projects for a particular domain or to the organizational unit that brings together researchers from various disciplines to carry out the projects. When several research organizations are involved, the program constitutes a national research network for the domain or subsector in question (Janssen & Kissi 1997). 2. A group of related projects managed in a coordinated way. Programs usually include an element of ongoing activity (Project Management Institute 1996).

Program plan: 1. A plan in which program research content is broadly defined for the medium to long term. Program plans also identify the resources needed to implement the program, mainly human resources (number of researchers and mix of disciplines), special equipment if any, and an indication of the amount of funding required for the period considered. Funding is based on a norm for operating cost per researcher although this is country specific. 2. A term that refers to all of the following: benefits management plans, risk management plan, transition plan, project portfolio plan and design management plan (Welcom Glossary).

Project: 1. A temporary endeavor undertaken to create a unique product or service (Project Management Institute 1996). 2. A coherent set of experiments or studies necessary to accomplish a goal. A research project may aim to develop a technology or methodology and is often executed by a group of researchers with different disciplinary backgrounds. Research projects are the buildings blocks of the program (based on Janssen and Kissi 1997). 3. A set of research activities designed to achieve specific objectives within a specified period of time. A research project is a group of interrelated research activities or experiments that share a rationale, objectives, plan or action, schedule for completion, budget, inputs, outputs, and intended beneficiaries.

Project appraisal: The discipline of calculating the technical and economic viability of a project, usually by a potential donor.

Project management: 1. Application of knowledge, skills, tools, and techniques to project activities in order to meet or exceed stakeholders needs and expectations from a project. This involves balancing competing demands among (a) scope, time, cost, and quality, (b) stakeholders with differing needs and expectations, (c) identified requirements (needs) and unidentified requirements (expectations). 2. The term project management is sometimes used to describe an organizational approach to the management of ongoing operations. This approach, more properly called management by projects, treats many aspects of ongoing operations as projects in order to apply project management to them (Project Management Institute 1996). 3. Approach used to manage work with the constraints of time, cost, and performance targets (Welcom Glossary).

Project plan: 1. A formal, approved document or design used to guide project execution. The primary uses of the project plan are to document planning assumptions and decisions, to facilitate communication among stakeholders, and to document approved scope, cost, and schedule baselines. A project plan may be summary or detailed (Project Management Institute 1996). 2. A document for management purposes that gives the basics of a project in terms of its objectives, justification, and how the objectives are to be achieved. This document is used as a record of decisions and a means of communication among stakeholders (Welcom Glossary). See also *plan*.

Project planning: Project planning is a systematic and integrated management approach to identifying and preparing a plan to resolve a "problem" identified within a certain field. Research project planning systematically identifies and responds to a need for research or its core "problem". Beneficiaries and stakeholders are also identified. Planning anticipates that the research project will develop an optimum solution to this need, including elaboration of the best means by which the solution may be achieved.

Research policy: 1. An (agricultural) research policy is a consensus statement – based on a nation's underlying philosophy, values, and societal aspirations – of what categories of knowledge and technologies to generate and diffuse, how and by whom they are to be generated in the most socially cost effective manner, to achieve sustainable agricultural development through the realization of stated agricultural research policy objectives.

Research results: Encompasses both outcomes and outputs. Results expected from a certain research activity (e.g., a project) can take the form of new information, or the development or improvement of a new technology. See also *impact*, *outcomes*, and *outputs*.

Research stakeholder: 1. The groups whose interests are likely to be affected by the research activities or, conversely, whose activities will affect the research system. 2. Individuals and organizations that are involved in or may be affected by project activities (Project Management Institute 1996). 3. Stakeholders are the people who have a vested interest in the outcome of the project (Welcom Glossary).

Research year Twelve months of research activities. The time a researcher devotes to "research" over the course of a calendar year is, in the strict sense of the word, less than 12 months. A research year may therefore cover more than one calendar year of a researcher's time. A research year can extend over a calendar year, financial year or cropping year.

Resource: An item required to accomplish an activity. Resources can be people, equipment, facilities, funding, or anything else needed to perform the work of a project. (Welcom Glossary)

Scenario: 1. A scenario is a description of a vision of the future state of a system. It is based on an assessment of the environment, of the forces of change at work, and the likely interaction between system variables in the progression from current conditions to a future state (Collion 1989). 2. Alternative scenarios describe the future conditions in which an organization (or system) may have to operate, as defined by sets of distinct hypotheses on the key variables that affect the development of the organization. Scenario development is a means for generating relevant information about an uncertain future.

Simulation model: Simulation models constitute a class of symbolic models, which are abstract representations of particular facets of reality that are built for specific purposes. Computer-based and essentially mathematical, complete flexibility is allowed as to its underlying structures. A good simulation model operates on input data to produce output data by mimicking particular processes and parts of reality that are of interest to the user.

Stakeholders: See *research stakeholder*.

Strategic planning: 1.A process by which an organization builds a vision of its future and develops the necessary structure, resources, procedures, and operations to achieve it. 2. A disciplined effort to produce fundamental decisions and actions that shape and guide what an organization (or other entity) is, what it does, and why it does it (Bryson 1995).

Strategy: 1. A course of action, chosen to reach a long-term vision or goal. 2. The pattern of objectives, purposes, or goals and major policies and plans for achieving those goals, stated in such a way as to define what business the company or organization is in or is to be in and the kind of company or organization it is or is to be (based on Hamermesh 1983).

Subprogram: A research program can be divided into subprograms. When a program covers several products or commodities, there is often a subprogram for each. For example, a cereals program may have subprograms for wheat and barley. Alternatively, subprograms may correspond to agroecological zones if there is sufficient diversity to justify the pursuit of separate research strategies (Janssen and Kissi 1997).

Support project: During program formulation, the planning group may identify certain projects to be carried out in addition to those research projects that are supposed to lead directly to new or improved technologies. A support project is one that generates information useful to researchers or development authorities. For example, the information may improve researchers' understanding of the physical or socioeconomic environment, allowing for better targeting of technologies. Or it may increase the probability of success of another research project by enhancing control of intervening factors (Janssen and Kissi 1997).

Target group: The intended recipients or consumers (based on Scriven 1991).

Task: A basic component of a research activity. Tasks, when combined, make it possible to achieve the intended result of the activity. In principle, the researcher in charge of the activity carries out all tasks of a research activity. However, in practice one or more tasks may be handled by a technician under the researcher's supervision or by another researcher.

Team, group, or committee: A team, group, or committee is made up of two or more people working interdependently toward a common goal and a shared reward (based on Welcom Glossary).

Technological innovation: 1. Technological innovation occurs in a well defined sequence. First there is the scientific discovery of the principle that precedes the innovation. The next step is the invention, which is the first working model of the innovation resulting from the discovery. The innovation is the first feasible demonstration of the invention. This usually occurs when the good is ready to be mass produced, sold commercially or made available through extension (based on Dunphy et al 1997). 2. Innovation is the process of creating something new that has significant value to an individual, a group, an organization, an industry, or a society (Higgins 1995).

Technology dissemination: 1. The (active) diffusion of a technology, for example, through an extension agency (based on Webster). 2. Broad concept that includes the diffusion of agricultural innovations to the farmer and the provision of prerequisites needed to make adoption possible (based on Arnon 1989).

Training Plan: Training planning is the process of designating knowledge, skills, and attitudes to be developed by the research organization through its training programs. Purposeful development can be more or less formal, ranging from classroom instruction to coaching during certain job assignments. Training planning in agricultural research is the process of choosing goals for formal training (degree and nondegree) and specifying what, where, and how training programs should be undertaken by which staff.

Users of research results: Direct and indirect users of technologies developed by research. Direct users include the various types of crop and livestock producers, agroindustries, traders, and scientists. Indirect users are those who incorporate research results into their activities: decision makers, development authorities, extension managers and so on (based on Janssen and Kissi 1997). See also *beneficiaries* and *target groups*.

References

Alston, J. M., G. W. Norton, P. G. Pardey. 1998. Science under Scarcity: Principles and Practice for Agricultural Research Evaluation and Priority Setting. Wallingford, UK: CAB International.

Arnon, I. 1989. Agricultural Research and Technology Transfer. Essex, UK: Elsevier Science Publishers.

Bryson, J. M. 1995. Strategic Planning for Public and Non-profit Organizations San Francisco, Calif.: Jossey-Bass.

Dunphy, S., P. A. Herbig, and F. A. Palumbo. 1997. Structure and innovation. In *The Innovation Challenge* by D. E. Hussey, pp. 195–219. West Sussex, UK: John Wiley & Sons.

Collion, M. -H. 1989. Strategic Planning for National Agricultural Research Systems: An Overview. Working Paper 26. Online: April 1999. (http://www.fao.org/news/1997/introG-e.html)

Hamermesh, R.G. 1983. Introduction. In *Strategic management. Harvard Business Review Executive Book.* West Sussex, UK: John Wiley & Sons.

Hartwich, F. 1998. A Cumulative Indicator for Measuring Agricultural Research Performance. Discussion Paper No. 98-1. The Hague: International Service for National Agricutural Research.

Henry, P. 1985. Evaluation: a systematic approach. Sage publications, California.

Higgins, J. M. 1995. Innovate or Evaporate: Test and Improve Your Organization's IQ – Its Innovation Quotient. Winter Park: New Management Publishing Company.

Horton, D., P. G. Ballantyne, W. Peterson, and B. Uribe. 1993. Monitoring and Evaluating Agricultural Research: A Sourcebook Wallingford, UK: CAB International.

Horton, D., R. Mackay, A. Anders, L. Dupleich. 2000. Evaluating Capacity Development in Planning, Monitoring and Evaluation: A Case from Agricultural Research. ISNAR Research Report 17. The Hague: International Service for National Agricutural Research.

ISNAR. 1997. Strengthening the role of universities in the national agricultural research systems in sub-Saharan Africa. Cotonou Benin, 17-21 November 1997. The Hague: International Service for National Agricutural Research.

Jain, H. K. 1990. Research Planning and Programming – Some Considerations of Terminology and Definition. Mimeo. The Hague: International Service for National Agricutural Research.

Janssen, W. and A. Kissi. 1997. Planning and Priority Setting for Regional Research. Research Management Guidelines 4. The Hague: International Service for National Agricutural Research.

Kissi, A. n.d. Planning Vocabulary in Agricultural Research. Mimeo. The Hague: International Service for National Agricutural Research.

Kronen, M. 1996. Logical Framework Manual. Suva, Fiji: Pacific Regional Agricultural Programme Logical Framework Training Kit.

Lusthaus, C., G. Anderson, E. Murphy. 1995. Institutional Assessment: A Framework for Strengthening Organizational Capacity for IDRC's Research Partners. Ottawa: International Development Research Centre.

Mills, B. (ed). 1998. Agricultural Research Priority Setting, Information Investments for the Improved Use of Research Resources. The Hague: International Service for National Agricutural Research.

NORAD. 1990. The Logical Framework Approach – Handbook for Objectives-Oriented Project Planning. Oslo: Norwegian Agency for Development Cooperation

OECD (Organisation for Economic Co-operation and Development). 1999. Online. http://www.oecd.org//puma/mgmtres/pac/pubs/evaluation.pdf.

Principa Cybernetica Web. 1999. Online. http://pespmc1.vub.ac.be/asc/DELPHI_METHO.html)

Project Management Institute. 1996. http://www.pmi.org/publictn/pmboktoc.htm.

Retzlaff, R. H. 1992. Preparing a Medium-term Plan for a National Agricultural Research Organization (NARO). Discussion Paper 92-1. The Hague: International Service for National Agricutural Research.

Sartorius, R. 1996. The third generation logical framework approach: dynamic management for agricultural research projects. *European Journal of Agricultural Education and Extension*, 2 (4): 49–62.

Scriven, M. 1991. Evaluation Thesaurus. California: Sage Publications.

Staveren, J.M. van, D. B. van Dusseldorp. 1980. Framework for Regional Planning in Developing Countries. Wageningen, The Netherlands: International Institute for land Reclamation and Improvement.

Webster. 1999. Online dictionary. http://www.m-w.com/netdict.htm.

Welcom Glossary. 1999. http://www.welcom.com/library/glossary/index.htm.

UNICEF. 1990: A UNICEF Guide for Monitoring and Evaluation: Making a Difference? The Division of Evaluation, Policy and Planning.
Also online at http://www.unicef.org/reseval.

Websites

This section offers a selection of organizations that have websites with information relevant to planning. No reference is made to specific pages, as these links become outdated rapidly. For updates, visit ISNAR's website (http://www.cgiar.org/isnar).

Strategic Planning

National Science Foundation	http://www.nsf.gov
De Wit & Meyer	http://www.dewit-meyer.com
APQC	http://www.apqc.com

Science and Technology Foresight

APEC STF	http://www.nstda.or.th/apec
Australian STF	http://www.astec.gov.au
IFPRI Vision 2020	http://www.cgiar.org/ifpri
Dutch National Council for Agricultural Research	http://www.agro.nl/nrlo
New Zealand Foresight Project	http://www.morst.govt.nz/foresight
UK Foresight Project	http://www.foresight.gov.uk

Training Planning

Economic Development Institute, World Bank	http://www.worldbank.org/html/edi
ISNAR	http://www.cgiar.org/isnar
International Society for Performance Improvement	http://www.ispi.org
Society for Human Resource Management	http://www.shrm.org

Planning, Monitoring, and Evaluation

OECD Working Party on Aid Evaluation	http://www.oecd.org/dac/evaluation
Mande News	http://www.mande.co.uk
Eldis participatory monitoring and evaluation	http://www.ids.ac.uk
PREVAL (International Fund for Agricultural Development and Inter-American Institute for Cooperation on Agriculture)	http://www.fidamerica.cl
IUCN monitoring and evaluation initiative	http://www.iucn.org
The American Evaluation Association	http://www.eval.org

Gender Analysis

International Development Research Centre, Gender and Sustainable Development Unit	http://www.idrc.ca/socdev
World Bank (Participation Sourcebook)	http://www.worldbank.org/wbi
Food and Agriculture Organization of the United Nations (SEAGA)	http://www.fao.org/sd/seaga

Participatory Approaches

International Development Research Centre, Social Policy Assessment Research	http://www.idrc.ca/socdev
International Institute for Environment and Development	http://www.oneworld.org/iied

Geographic Information Systems

International Institute for Aerospace Survey and Earth Systems	http://www.itc.nl

About the Authors

Gary Alex is an agricultural development specialist working as a consultant to the World Bank. His main interests and experience are in agricultural technology systems – research, extension, and education. He has over 30 years' experience with agricultural development programs in Asia, Latin America, and Africa.

Jacqueline Ashby is a rural sociologist with a PhD from Cornell University. She has contributed extensively to the development of particpatory methods for agricultural research. Currently she is director of the Natural Resource Management division of the International Center for Tropical Agriculture (CIAT) and leader of the CGIAR system-wide initiative on participatory research and gender analysis.

Henning Baur is a senior research officer at the International Service for National Agricultural Research (ISNAR). He has worked as an agricultural economist and research management advisor in Africa and the Middle East. His major interest is in organizational development and the optimization of innovation processes.

Thomas Braunschweig is a research fellow at ISNAR. He is also associated with the Swiss Federal Institute of Technology (ETH) in Zurich. His main interest is in decision-making processes for public-sector research, particularly in setting priorities for agricultural biotechnology.

Hilarion Bruneau is a financial management specialist at the World Bank, working with World Bank Group clients, borrowers, and World Bank country and project teams to ensure that loan disbursement is efficient and in support of targeted development objectives. His focus is mainly on Indonesia in the social and agricultural sectors.

Edwin Brush has 25 years' international experience in the strategic development and management of human resources for science and technology. He has advised government agencies and private industry about human resources for educational systems, environmental management, agricultural research, and power and transportation systems.

Derek Barley is principal economist and agricultural research specialist in the Rural Development Department of the World Bank. His main interests are in agricultural research policy and management.

Olga Capo Iturrieta is an industrial engineer who works in the Planning Unit of Instituto Nacional de Investigaciones Agropecuarias (INIA), Chile. She works in developing methodological tools to assist in designing, monitoring, and evaluating INIA's agricultural research projects and the subsequent analysis of project performance data to guide institute management.

Robert Chambers is a research associate at the Institute of Development Studies at the University of Sussex. His main current interest is participatory approaches, methods, and behaviors in development.

Marie-Hélène Collion is presently a senior agricultural services specialist at the World Bank, working on research and extension projects and support to producers' organizations in West Africa. She was previously senior officer at ISNAR working primarily on agricultural research planning and priority setting.

Rudolf Contant is a geneticist and agronomist from Wageningen University. He was professor and head of the Department of Applied Plant Sciences at the University of Nairobi and developed the Association of Faculties of Agriculture in Africa in the 1970s. He is a senior research officer at ISNAR, focusing on priority setting and strategic planning.

Luis Dupleich is an economist in the Policy Analysis Unit of the Government of Bolivia. In 1998 and 1999 he worked for ISNAR on issues of planning, monitoring, and evaluation. In 1997 he earned a master's degree in agricultural and rural development from the Institute of Social Studies in the Netherlands.

Jörg Edsen graduated from the University of Hohenheim in Germany as an agricultural economist – tropical agronomist. He was employed at ISNAR for six years as a research associate. He is now based in Uganda and works as a freelance consultant in agricultural research management.

Cesar A. Falconi is an economist at the Inter-American Development Bank (IDB). Previously he worked as a research officer at ISNAR. His main interests are policy and management issues in biotechnology research, intellectual property rights, and public and private partnerships.

Silvia Galvez Anastassiou is a Chilean agricultural engineer. She has worked mainly in agricultural technology transfer and in agricultural research planning, monitoring, and evaluation. In her position as Director of the Planning Unit in Instituto Nacional de Investigaciones Agropecuarias (INIA), Chile she is responsible for developing and maintaining an integrated PM&E system.

Govert Gijsbers is a research officer at ISNAR. His main interests concern issues of institutional development and governance of agricultural research and innovation.

Helen Hambly Odame is an associate research officer in the training and capacity-building unit at ISNAR. She is a graduate of the Faculty of Environmental Studies, York University, Canada, with research experience in sub-Saharan Africa.

Vanessa Henman-Bainbridge, MSc Forestry, gained experience in farmer participatory research while working for CARE Development in a conservation project in south-west Uganda. She has since worked as a research assistant at the Institute of Development Studies at the University of Sussex and is currently looking after her newborn son.

Brigitte Holzner is a psychologist and development sociologist, and works as senior lecturer at the Institute of Social Studies in The Hague. Her main interests are gender issues and planning methodologies with respect to rural development and reproductive health.

Douglas Horton is an economist and evaluation specialist employed at ISNAR. Previously, for 15 years, he was head of the social science department of the International Potato Center in Peru. He received his bachelor's and master's degrees in agricultural economics from the University of Illinois and his PhD in economics from Cornell University.

Willem Janssen is an agricultural economist with a doctoral degree from Wageningen Agricultural University. He is a senior research officer at ISNAR. His work concerns institutional innovation (especially concerning new technological demands), priority-setting, financing, and policy-formulation methodologies.

Bruce B. Johnson is a research professor and project supervisor at the Management Institute Foundation of the University of São Paulo, Brazil. His main interests are technology foresight, research strategy, and technology management.

Ali Kissi is an agronomist and inspector general from Morocco. Since 1990, he has been involved in the development of program planning methods and has applied such methods in many sub-Saharan countries. He has also been involved in the development of project planning tools, monitoring, evaluation, and auditing methods and assessment of institutional structure.

Ronald Mackay is a professor of education at Concordia University, Montreal with a special interest in development project management and evaluation and total land management. He was a senior research fellow at ISNAR from January to December 1997.

Gerdien Meijerink is a socioeconomist who graduated from Wageningen Agricultural University in 1994. After working as a consultant in the area of natural resource management, she joined ISNAR in 1996, where she has been working on various subjects, including natural resource management and gender.

Maria Lúcia D'Apice Paez is a researcher at Embrapa (Brazilian Corporation of Agricultural Research, Brazil). She earned her PhD in agricultural and resource economics from Oregon State University. She previously worked at the Agricultural Research Institute (IEA, São Paulo State). Recently, she's returned from a post-doc at the Department of Economics, University of São Paulo, Agribusiness Program (PENSA).

Paul Perrault is a senior research officer in ISNAR's information and new technologies program where he leads a distance-training project on agricultural research management.

Previously, he led ISNAR's program on research management. His main research interest is organizational change.

Warren Peterson holds a PhD in anthropology and masters' degrees in extension sciences and agricultural economics. He started to work with national agricultural research systems in 1985 while employed by the International Agricultural Program of the College of Agriculture of the University of Illinois. He joined ISNAR in 1990. Since that time he has assisted more than 25 countries in developing their agricultural research capabilities, especially in the fields of planning, monitoring, and evaluation.

Louise Sperling is a social scientist with over 20 years of field experience in Africa and Asia working with farming communities, national agricultural research systems, and a wide variety of intermediary organizations. At present she facilitates a working group of participatory plant breeding under the umbrella of the CGIAR Systemwide Program on Participatory Research and Gender Analysis (PRGA). The group helps plant breeders make their work more responsive to end-user needs (especially in marginal areas and for women and poor farmers) and strengthens farmers' own local systems of plant breeding and seed supply maintenance.

Hans Rutten is an agricultural economist from the Netherlands, graduated from Wageningen Agricultural University (WAU). He has worked at the WAU, the Agricultural Economics Research Institute (LEI-DLO), and as foresight expert at the National Council for Agricultural Research (NRLO). He now works at the Netherlands' Ministry of Agriculture, Nature Management and Fisheries in the Strategic Policies Division.

José de Souza Silva is a Brazilian sociologist of science and technology working with ISNAR. He managers the ISNAR's "new paradigm" project for Latin America, which aims to strengthen the institutional sustainability of agricultural science and technology organizations.

Steven R. Tabor works as an economic advisor to the governments of Sri Lanka and Indonesia. He has a PhD in economics from the Vrije Universiteit of Amsterdam. Previously he worked at ISNAR on research policy and finance issues.

Philip Thornton is an agricultural systems analyst and project coordinator at the International Livestock Research Institute (ILRI) in Nairobi, Kenya. He works in systems analysis and impact assessment. He is currently involved in modeling weather, land-use change, and crop-livestock interactions in smallholder systems in the tropics and subtropics.

Jaime Tola is a senior research fellow at ISNAR with practical experience on institutional management in the Andean region and in Southern Africa.

Carlos Valverde received his MSc in soil science at Purdue University and a PhD in soil science management from Michigan State University. He was instrumental in the creation and development of the National Agrarian Research Institute (INIA) in Peru.

Valverde joined ISNAR in 1983 as a senior research fellow, becoming a senior research officer in 1985, and regional coordinator for Latin America and the Caribbean in 1992. In early 1999, he retired from ISNAR. Presently Valverde operates as an ISNAR Associate.

Richard Vernon is a tropical agriculturalist and information systems specialist. He is the author of several papers in tropical agriculture and of the book "Information Systems for Agricultural Research Management," ISNAR and Technical Centre for Agricultural and Rural Cooperation (CTA) (in press).

Acronyms

AHP	analytic hierarchy process
APAARI	Asia-Pacific Association of Agricultural Research Institutions (Thailand)
APEC	Asia-Pacific Economic Cooperation
ARC	Agricultural Research Council (South Africa)
ASARECA	Association for Strengthening Agricultural Research in Eastern and Central Africa
CAMASE	Concerted Action for the development and testing of quantitative Methods for research on Agricultural Systems and the Environment
CATIE	Centro Agronómico Tropical de Investigación y Enseñanza (Costa Rica)
CBA	cost-benefit analysis
CERAAS	Centre d'Etude Régional pour l'Amélioration de l'Adaptation à la Sécheresse
CGIAR	Consultative Group on International Agricultural Research
CIP	Centro Internacional de la Papa
CIRAD	Centre de Coopération Internationale en Recherche Agronomique pour le Développement (France)
CLAN	Cereal and Legumes Asia Network
CORAF	Conseil Ouest et Centre Africain pour la Recherche et le Développement Agricoles
CPM	critical path management
DSSAT	Decision Support System for Agrotechnology Transfer
Embrapa	Empresa Brasileira de Pesquisa Agropecuária (Brazil)
ET	evapotranspiration
FAO	Food and Agriculture Organization of the United Nations
FONTAGRO	Fondo Regional de Tecnología Agropecuaria
GIS	geographic information systems
GMOs	genetically modified organisms
GPRA	Government Performance and Results Act
GTZ	Gesellschaft für Technische Zusammenarbeit (Germany)
ICASA	International Consortium for Agricultural Systems Applications
ICRAF	International Center for Research in Agroforestry
IICA	Inter-American Institute for Cooperation on Agriculture
IITA	International Institute of Tropical Agriculture
ILRI	International Livestock Research Institute

IMF	International Monetary Fund
INIA	Instituto Nacional de Investigación Agraria (Uruguay)
INIA	Instituto (Nacional) de Investigaciones Agropecuarias (Chile)
INIAP	Instituto Nacional Autónomo de Investigaciones Agropecuarias (Ecuador)
INRA	Institut National de Recherche Agronomique (Morocco)
INRAB	Institut National des Recherches Agricoles du Bénin
INSAH	Institut du Sahel
INTSORMIL	International Sorghum and Millet Collaborative Research Support Program (USA)
IPM	integrated pest management
IPRs	intellectual property rights
ISNAR	International Service for National Agricultural Research
ISRA	Institut Sénégalais de Recherches Agricoles
ITC	International Institute for Aerospace Survey and Earth Sciences
KARI	Kenya Agricultural Research Institute
KSAs	knowledge, skills, and attitudes
MERCOSUR	Mercado Común del Cono Sur
MIS	management information system
MOA	Ministry of Agriculture (Palestine)
NPV	net present value
NRLO	National Council for Agricultural Research (The Netherlands)
NRM	natural resource management
NSF	National Science Foundation (USA)
OECD	Organisation for Economic Co-operation and Development
PA	Palestinian Authority
PERT	Program Evaluation and Review Technique
PM&E	planning, monitoring, and evaluation
PPBS	planning, programming, and budget system
PRA	participatory rural appraisal
PROCIs	Programas Cooperativos de Investigación y Transferencia de Tecnología Agrícola (Latin America)
RRA	rapid rural appraisal
S&T	science and technology
SACCAR	Southern African Centre for Co-operation in Agricultural and Natural Resources Research and Training
SINCITA	National Agricultural Science and Technology System (Cuba)
SINGER	System-wide Information Network for Genetic Resources (CGIAR)
STF	Science and Technology Foresight

SWOT	strengths, weaknesses, opportunities, and threats
USAID	US Agency for International Development (United States)
WARDA	West Africa Rice Development Association
WCASRN	West and Central African Sorghum Research Network
ZOPP	Zielorientierte Projekt-Planung (GTZ)

Index